Conventional and Alternative Power Generation

Conventional and Alternative Power Generation

Thermodynamics, Mitigation and Sustainability

Neil Packer and Tarik Al-Shemmeri

Registered Office
John Wiley & Sons Ltd, The Atrium, Southern Gate, Chichester, West Sussex, PO19 8SQ, UK

Editorial Offices
9600 Garsington Road, Oxford, OX4 2DQ, UK
The Atrium, Southern Gate, Chichester, West Sussex, PO19 8SQ, UK

For details of our global editorial offices, customer services, and more information about Wiley products visit us at www.wiley.com.

Wiley also publishes its books in a variety of electronic formats and by print-on-demand. Some content that appears in standard print versions of this book may not be available in other formats.

Library of Congress Cataloging-in-Publication Data

Names: Packer, Neil, author. | Al-Shemmeri, Tarik, author.
Title: Conventional and alternative power generation : thermodynamics, mitigation and sustainability / Neil Packer, Prof. Tarik Al-Shemmeri.
Description: 1 edition. | Chichester, UK ; Hoboken, NJ : John Wiley & Sons, 2018. | Includes bibliographical references and index. |
Identifiers: LCCN 2018006236 (print) | LCCN 2018012068 (ebook) | ISBN 9781119479376 (pdf) | ISBN 9781119479406 (epub) | ISBN 9781119479352 (cloth)
Subjects: LCSH: Electric power production. | Renewable energy sources. | Thermodynamics.
Classification: LCC TK1001 (ebook) | LCC TK1001 .P325 2018 (print) | DDC 621.31/21–dc23
LC record available at https://lccn.loc.gov/2018006236

Cover Design: Wiley
Cover Images: © chinaface/iStockphoto;
© westcowboy/iStockphoto;
© Diyana Dimitrova/Shutterstock

Set in 10/12pt Warnock Pro by SPi Global, Chennai, India

10 9 8 7 6 5 4 3 2 1

Contents

Preface

Thermodynamics, often translated as 'movement of heat', is simply the science of energy and work. Energy itself is described as the capacity to do work.

French steam engineer Nicolas Leonard Sadi Carnot, who was well aware that the realization of water power is a function of water level or head difference across a turbine, suggested in 1824 that capacity for work and power across a *heat engine* would be dependent on the prevailing temperature difference.

Between 1840 and 1850, British scientist and inventor James Joule investigated the nature of work in a range of forms, for example, electrical current, gas compression and the stirring of a liquid. He concluded from his work that 'lost' mechanical energy would express itself as heat, for example, friction, air resistance etc., and hence spoke of the *mechanical equivalent of heat*.

In 1847, German physicist, Hermann Von Helmholtz first postulated the principle of energy accountancy and energy conservation. In 1849, British physicist William Thomson (later Lord Kelvin) is thought to have coined the term *Thermodynamics* to describe the subject of energy study, and the Helmholtz principle became enshrined as the *First law of thermodynamics*.

In 1850, German physicist Rudolf Julius Emmanuel Clausius used the term *entropy* to describe *non-useful heat* and proposed that, in universal terms, entropy increase is a natural, spontaneous process, leading to the development of a *Second law of thermodynamics*. This can be stated in several ways but perhaps the simplest is that it is not possible for an engine operating in a cycle to convert heat into work with 100% efficiency.

Civilizations are often judged on their cultural legacy, described in terms of their contribution to architecture, art and literature, and its spread across the globe.

It could be argued that the current manifestation of human civilization will be judged on the legacy of its technological ingenuity and, in particular, its endeavours to supply energy to a rapidly expanding planetary population seeking ever-increasing standards of living.

The challenge is to make the most efficient use of energy sources and produce power at the minimum cost and least environmental impact. Failure to achieve this has global consequences in terms of an unwanted environmental legacy.

This book examines currently available conventional and renewable power-generation technologies and describes the allied pollution-control technologies associated with the alleviation of their environmental impact.

Neil Packer and Tarik Al-Shemmeri

Structure of the Book

Chapter 1

Flemish (modern-day Belgium) chemist Jan Baptista van Helmont first coined the terms *gas* and *vapour* in the 17th century. He related the classification to ambient temperature. Substances like oxygen, nitrogen and carbon dioxide are gases at ambient temperature whereas substances like water can only be gasified at an elevated temperature, making steam, a vapour.

In any heat engine, the transfer of energy from place to place is the job of the working fluid. Working fluids in heat engines are liquids, vapours and gases, and so Chapter 1 looks at some of the fundamental properties of these phases in relation to their energy content and introduces the reader to the use of standard property tables and charts.

Chapter 2

For about 100 years from the late 18th century, the reciprocating piston/cylinder and drive wheel steam engine dominated mechanical power production. However, in 1884, British engineer Charles Algernon Parsons changed all that when he conceived a new technology for accessing the power of steam. His revolutionary idea was to use nozzles to direct high-pressure, high-temperature steam jets onto a series of engineered blades connected at their roots to a shaft, thus causing the shaft to rotate. Originally, his idea was deployed in a marine transport application but it was not long before his steam turbine was connected to a stationary generator by the fledgling electrical power supply industry at the beginning of the 20th century.

However, vapour power generation comprises a number of processes and technologies in addition to the turbine, for example, a boiler, condenser, pump etc., making up a cycle, and so Chapter 2 progressively introduces the reader to complex vapour power cycles, enabling the calculation of fundamental performance parameters such as cycle efficiency, SSC etc.

Chapter 3

A power-generation system employing non-condensable gases is not a new idea.

In 1816, a member of the Scottish clergy, Reverend Robert Stirling, proposed a heat engine based on the sequential heating and cooling of air. However, at the time, the design was not a great success because of the limitations of contemporary material science knowledge.

Gas reciprocating engines for small motive power applications have been in development since 1860, when Frenchman Jean Joseph Etienne Lenoir exhibited his horizontal, double-acting, single-cylinder, non-compression machine running on coal gas and air. Although earlier attempts were made, German inventor Nicolaus August Otto is credited with the first compressed-gas, electrically ignited, four-stroke engine patented in 1866. (French-born) German engineer Rudolf Christian Karl Diesel patented an engine in 1892 that did not rely on a spark for ignition but instead achieved a flash point temperature for the fuel by compression alone, making it suitable for use with liquid fuels.

Proposals and patents for the gas turbine can be traced back to the mid-18th century. However, the production of the first practical industrial gas turbine power plant is credited to the Swiss company Brown Boveri in 1939. Since that time, many improvements have been proposed, including multiple shafts, exhaust gas recuperation, intercooling and reheating and closed and combined cycles.

In the modern era, gas cycles in stationary power generation also play an increasingly important role in base load, decentralized, standby and peak lopping applications. Chapter 3 covers gas power generation cycles in both rotary compressor/turbine schemes and displacement engines.

Chapter 4

In distant historical times, mechanical power had to be supplied by man's own muscles or by those of his animals. Rendered animals and plants could also be the source of oils that would burn for illumination purposes. For most of our history, however, wood has been our major source of fuel. There is evidence to suggest that coal was used as a fuel from about the 1200s onwards. At the beginning of the 17th century, coal was discovered to be a potential source of derived fuels if heated in the absence of air.

In 1859, American Edwin Laurentine Drake thought that a naturally occurring, high-density, inflammable liquid that was found in association with shale deposits might have an economic value for lighting purposes, and he drilled the first oil well in Titusville, western Pennsylvania, USA. By the turn of the century, this *crude* oil was being distilled and reformed to produce a range of liquid and vapour fuels. In fact, it was soon discovered that *natural* gaseous resources could often be found underground *in situ* with oil deposits.

Our entire civilization now depends on these fossil fuels (coal, oil and gas) and they are used extensively as the energy source in the previously described power cycles. Chapter 4 explores their properties, the chemical changes taking place during their burning or combustion with air, the prediction of the energy released and the nature of some gaseous emissions associated with their use.

Chapter 5

Liquid and solid fuels tend to have mineral content as part of their composition, and so the consumption of some fuels, for example, coal, diesel and biomass, has associated with it the significant generation of potentially harmful particulate matter.

The size of this particulate matter tends to be on the micron scale, making it particularly harmful if inhaled. The World Health Organization suggests annual and 24-hour concentration exposure limits for the most dangerous 10 μm and 2.5 μm diameter particles.

In 1851, Irish physicist and mathematician, George Gabriel Stokes provided a model predicting the resulting velocity for a small particle falling under the effect of gravity. This simple understanding underpins much of Chapter 5, which looks at the nature of particulates in a fluid stream and describes the theory and operation of a range of pollution-control devices to capture them and alleviate the problem.

Chapter 6

Our planetary heat exchange with space is dependent on solar input, surface reflectivity and the composition of our atmosphere. The heat balance determines our planet's average temperature. The emissions associated with our fossil-fuel-based industrialization have dramatically altered the earth's atmosphere and hence its equilibrium, resulting in a prediction of a significant increase in average temperature. At the end of the 19th century, Swedish scientist and climate modeller Svante Arrenius first postulated a link between increased atmospheric carbon dioxide concentration from fossil fuel use and a rise in global temperature. This effect is already in evidence in the second decade of the 21st century. There are a number of *global warming gases*, but the principal emission associated with this change is carbon dioxide, and Chapter 6 focuses on this aspect. Carbon dioxide is usually found in a combination with other emissions, and so properties of gas mixtures are introduced along with a thermodynamic analysis of gas separation. The chapter goes on to look at practical separation techniques as well as some proposed storage solutions to the problem.

Chapter 7

In fossil fuel power generation, whatever remains of the products of combustion after filtration is transferred to a stack or chimney for release to the atmosphere. Its atmospheric dispersion is, essentially, an example of the diffusion spreading of one substance in another along a concentration gradient. The laws governing the resulting diffusive flux and concentration field were laid down by German physicist and physiologist Adolf Fick in 1855. For a gaseous stack emission, ambient pressure, temperature and wind velocities would modify this diffusion, and so having an understanding of atmospheric phenomena and plume characteristics is key to acceptable and legal dispersal of emissions. This understanding was provided in the 1950s and 1960s by F. A. Gifford and British scientist Frank Pasquil. Chapter 7 looks at the principles of simple dispersal modelling, enabling the reader to predict air pollutant concentrations downwind of a stack.

Chapter 8

Generating energy from fossil fuels is ultimately unsustainable, as they are a finite resource and raise global warming issues. Sustainable power generation requires the deployment of renewable energy sources such as the sun, the wind and biomass. Although, strictly speaking, not renewable, some countries also consider the use of nuclear fuel as part of the solution. Of course, the switch cannot be achieved overnight but forward-looking countries around the world are already moving slowly towards this objective. Renewable sources tend to be intermittent and so energy storage will be required with their use. Chapter 8 reviews a range of renewable energy technologies and some measures by which to match their supply with a varying demand.

Notation

A text covering a broad range of topics will unavoidably require a large nomenclature. In general, parameters are introduced along with their units in the text. However, there are a few cases where the reuse of a symbol in a different context has become necessary to maintain coherence. The listings below highlight the reuse of a parameter by indicating, in brackets, the chapter of its subsequent reoccurrence.

Occasionally, differentiation may require an extended subscript. For example, amb – ambient, comb – combustion, gen – generated etc. Notation such as this is self-evident and will not be included in the list below.

English Symbols

A	area	m^2
a	wave amplitude	m
BF	buoyancy flux parameter	m^4/s^3
C	capacitance	Farad
C_p	specific heat capacity at constant pressure	kJ/kg K
C_v	specific heat capacity at constant volume	kJ/kg K
c	concentration	kg/m^3
D_A	diffusion coefficient	m^2/s
d	diameter	m
d	distance (8)	m
E	electric potential	volts
EF	electric field	volts/m
F	force	N
F	Faraday's constant (8)	kJ/kmol V
f	frequency	Hertz
$G, \bar{g}$	Gibbs energy	kJ, kJ/kmol
g	gravitational acceleration	m/s^2
$H, h, \bar{h}$	enthalpy	kJ, kJ/kg, kJ/kmol
H, h	height (5, 6)	m
h	Planck's constant (8)	Js

I	current	amps
I_m	moment of inertia	kg m^2
I_s	insolation	kWh/annum
K_H	Henry's coefficient	Pa
L	length	m
L	self-inductance (8)	Henrys
M	molar mass	kg/kmol
$m, \dot{m}$	mass, mass flow rate	kg, kg/s
N	rotational speed	rpm
$n, \dot{n}$	number of moles, molar flow rate	kmol, kmol/s
P	pressure	Pa
P_A	permeance	m^3/m^{2}sPa
Q, q	heat content	kJ, kJ/kg
$\dot{Q}$	heat transfer	kW
q	electrical charge (5)	Coulombs
R_o	universal gas constant	kJ/kmol K
R	specific gas constant	kJ/kg K
R	electrical resistance (8)	Ohms
r	radius	m
$S, s, \bar{s}$	entropy	kJ/K, kJ/kg K, kJ/kmol K
S_A	solubility coefficient	m^3/m^3 m Pa
T	temperature	°C, K
T	wave period (8)	seconds
t	time	s
$U, u, \bar{u}$	internal energy	kJ, kJ/kg, kJ/kmol
$V, v, \bar{v}$	volume	m^3, m^3/kg, m^3/kmol
V	voltage (8)	volts
$\bar{V}$	velocity	m/s
$\dot{V}$	volume flow rate	m^3/s
W, w	work	kJ, kJ/kg
W	width (5)	m
$\dot{W}$	work transfer	kW
X	exergy	kJ/kg
z	height above a datum	m

Subscripts/Superscripts

a	air	
B	buoyancy	
b	boiler	
b	building (7)	
C	centrifugal	

c	compression
c	collection (5)
cr	critical
cw	cold water
D	drag
d	diameter
E	electrostatic
e	emission
f	liquid
f	formation (4)
f	friction (8)
fb	fibre
fg	liquid– gas condition
G	gravitational
g	gas or vapour
gb	gearbox
g-g	gas in gas
g-l	gas in liquid
H	high
I	intermediate
I	inertial (5)
i	isentropic
i	component (4)
L	low
L	liquid (5)
o	standard state
p	pump
p	process (3)
p	particle (5)
R	resultant
r	reaction (4)
r	relative (5)
s	steam
st	stack
t	turbine
th	thermal
ts	terminal settling
u	utilization
v	void
w	wind
x, y, z	co-ordinates

Greek Symbols

α	Seebeck coefficient	V/K
ε_o	permittivity of free space	C/Vm
η	efficiency	%
λ	wavelength	m
μ	dynamic viscosity	kg/ms
π	Peltier coefficient	V
ρ	density	kg/m^3
ρ	electrical resistivity (8)	Ω/m
σ	dispersion coefficient	m
ω	rotational speed	rad/s
Ψ, ψ	thermodynamic property	various

Dimensionless Numbers

A/F	air–fuel ratio
C_D	drag coefficient
CF	correction factor
ε	dielectric constant
ε_u	utilization factor (2)
f_f	fibre solid fraction
ϕ	equivalence ratio
γ	ratio of specific heats
K	acentric factor
k	isentropic condition index
k	geometric factor (8)
mf	mass fraction
n	expansion/compression index
p	wind elevation exponent
Re	Reynolds number
r_c	cut-off ratio
r_v	compression ratio
Stk	Stokes's number
WR	work ratio
x	dryness fraction
y	molar fraction
z	compressibility factor

1

Thermodynamic Systems

1.1 Overview

Thermodynamics is the science relating heat and work transfers and the associated changes in the properties of the working substance within a predefined working system. A thermodynamic system is one that is concerned with the generation of heat and/or work using a working fluid. In this chapter, thermodynamic system behaviour will be described and the changes in properties will be calculated during the different processes encountered in typical engineering applications.

Learning Outcomes

- To understand the basic units and properties of thermodynamic systems.
- To be able to apply the laws of thermodynamics to closed and open systems.
- To be able to apply the first law of thermodynamics and calculate the changes in properties during a process and a cycle.
- To be able to solve problems related to compression and expansion of steam and gases.

1.2 Thermodynamic System Definitions

A thermodynamic *system* comprises an amount of matter enclosed within a *boundary* separating it from the outside *surroundings*.

There are two types of thermodynamic system:

A closed system has a fixed mass and a flexible boundary.
An open system has a variable mass (or mass flow) and a fixed boundary.

1.3 Thermodynamic Properties

A thermodynamic property of a substance refers to any quantity whose changes are defined only by the end states and by the process. Examples are the pressure, volume

Conventional and Alternative Power Generation: Thermodynamics, Mitigation and Sustainability,
First Edition. Neil Packer and Tarik Al-Shemmeri.
© 2018 John Wiley & Sons Ltd. Published 2018 by John Wiley & Sons Ltd.

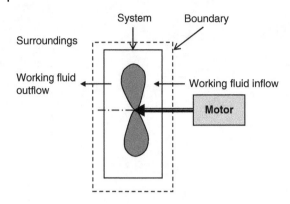

Figure 1.1 Thermodynamic system, boundary and surroundings.

and temperature of the working fluid in the system in Figure 1.1. In addition to these three properties, other thermodynamic properties include enthalpy, entropy and internal energy, which are all important in studying the behaviour of the working fluid in a power plant.

A list of the most common properties and associated terms is given below:

- *Pressure* (P) – The normal force exerted per unit area of the surface within the system. For engineering work, pressures are often measured with respect to atmospheric pressure rather than with respect to absolute vacuum.

  If a pressure gauge is calibrated to read zero at atmospheric pressure then the absolute pressure (P_{abs}) is given by:

  $$P_{abs} = P_{atm} + P_{gauge}$$

  In SI units, the derived unit for pressure is the pascal (Pa), where $1\ Pa = 1\ N/m^2$. This is very small for engineering purposes, so usually pressures are quoted in terms of kilopascals ($1\ kPa = 10^3\ Pa$), megapascals ($1\ MPa = 10^6\ Pa$) or bars ($1\ bar = 10^5\ Pa$).
- *Specific volume* (v) – For a system, the specific volume is the space occupied by a unit mass. The units of the specific volume are therefore m^3/kg.

  (Note that the term *specific* and lower-case letters are commonly used to denote thermodynamic property values *per kg of substance*.)
- *Temperature* (T) – Temperature is the degree of hotness or coldness of the system or the working fluid contained in the system. The absolute temperature of a body is defined relative to the temperature of ice at $0\ °C$.

  In SI units, the Kelvin scale is used, where $0\ °C \equiv 273.15\ K$.
- *Specific internal energy* (u) – The property of a system covering all forms of energy arising from the internal structure of its matter, for example, nuclear, molecular, vibrational etc. The units of specific internal energy are kJ/kg.
- *Specific enthalpy* (h) – An energy property of the system conveniently defined as the sum of the internal energy and flow work, i.e. $h = u + PV$ for a substance.

  The units of specific enthalpy are kJ/kg.
- *Specific entropy* (s) – Entropy refers to the microscopic disorder of the system. It represents the effect of *irreversibilities* due to friction and deviation from the ideal behaviour. Ideal processes are termed *isentropic*.

  The units of specific entropy are $kJ/kg\ K$.

- *Phase* – The condition of a substance described by terms such as *solid, liquid* or *gas* is known as its phase. Phase change occurs at constant temperature.
  Phase changes are described as follows:
  Condensation: gas (or vapour) to liquid
  Evaporation: liquid to gas (or vapour)
  Melting: solid to liquid
  Freezing: liquid to solid
  Sublimation: solid to gas
  Deposition: gas to solid.
- *Mixed phase* – A multi-phase condition, for example, ice + water, water + vapour etc.
- *Quality of a mixed phase or dryness fraction* (x) – The *dryness fraction* (no units, values 0–1) is defined as the ratio of the mass of pure vapour present to the total mass of a mixed phase. The *quality* of a mixed phase may be defined as the percentage dryness of the mixture.
- *Saturated state* – A saturated liquid is a state in which the dryness fraction is equal to zero. A saturated vapour has a quality of 100% or a dryness fraction of one.
- *Pure substance* – A pure substance is one which is homogeneous and chemically stable. Thus, it can be a single substance that is present in more than one phase, for example, liquid water and water vapour contained in a boiler (in the absence of any air or dissolved gases).
- *Triple point* – A state point in which all solid, liquid and vapour phases coexist in equilibrium.
- *Critical point* – A state point at and above which transitions between liquid and vapour phases are not clear.
- *Superheated vapour* – A gas is described as superheated when its temperature at a given pressure is greater than the phase change (or *saturated*) temperature at that pressure, i.e. the gas has been heated beyond its saturation temperature.
  The *degree of superheat* represents the difference between the actual temperature of a given vapour and the saturation temperature of the vapour at a given pressure.
- *Subcooled liquid (or compressed liquid)* – A liquid is described as undercooled or subcooled when its temperature at a given pressure is lower than the saturated temperature at that pressure, i.e. the liquid has been cooled below its saturation temperature.
  The *degree of subcool* represents the difference between the saturation temperature and the actual temperature of the liquid at a given pressure.
- *Specific heat capacity* (C) – An energy storage property dependent on temperature. There are two variants:
  Constant volume: C_v
  Constant pressure: C_p
  The ratio of the specific heat capacities, i.e. C_p/C_v, is a parameter that is important in isentropic processes. It represents the expansion/compression process index.

1.4 Thermodynamic Processes

A process describes the path by which the state of a system changes and some properties vary from their original values.

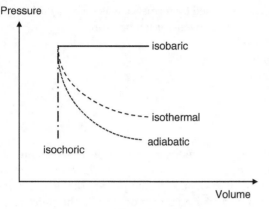

Figure 1.2 Thermodynamic processes.

In thermodynamics, the following types of processes are often encountered:

Adiabatic process: No heat transfers from or to the working fluid take place.
Isothermal process: No change in temperature of the working fluid takes place.
Isobaric process: No change in pressure of the working fluid takes place.
Isochoric process: No change in volume of the working fluid takes place.

These processes are illustrated on a pressure–volume basis in Figure 1.2.
Two other processes are of interest:

Isentropic process: No change in entropy of the fluid.
Isenthalpic process: No change in enthalpy of the fluid.

1.5   Formation of Steam and the State Diagrams

Consider the change in volume that occurs due to heating a unit mass of pure substance contained in a closed cylinder at constant pressure; take water (ice) as an example with an initial condition (temperature $-10\,°C$; pressure $100\,kN/m^2$). The change in specific volume corresponding to equilibrium states is illustrated by the successive piston displacements shown in Figure 1.3a and plotted on the temperature-specific volume diagram of state shown in Figure 1.3b. It can be seen that there is very little increase in the specific volume of ice up to the melting point, B. At the melting point, there is a small but significant contraction to point B′. This is not typical of all pure substances; many substances show a small expansion at the melting point. Further heating in the liquid phase results in a small expansion as the temperature is increased to the vaporization point, C. At this point, the liquid is said to be in the *saturated liquid condition.* Further heating causes the progressive conversion of liquid into vapour, with a large increase in specific volume while the temperature remains constant. The process of conversion from liquid to vapour is completed at point C′, which is known as the *saturated vapour condition.* Further heating gives rise to a gradual rise in temperature and specific volume in the vapour (or superheated steam) region, which is typified by point D.

Line ABB′CC′D represents a typical constant pressure line on the temperature-specific volume diagram of state in which the pressure lies above that at the triple point but well below the critical pressure. If the above process was carried out at a higher pressure, the large change in the specific volume at vaporization would be less pronounced. The abrupt change in specific volume disappears at the critical and higher pressures and a point of inflection occurs in the constant pressure line passing through the critical point.

At pressures above the critical value, there is no marked distinction in transition from the liquid to the vapour phase and it is usual to regard a substance as a gas at temperatures above the critical temperature.

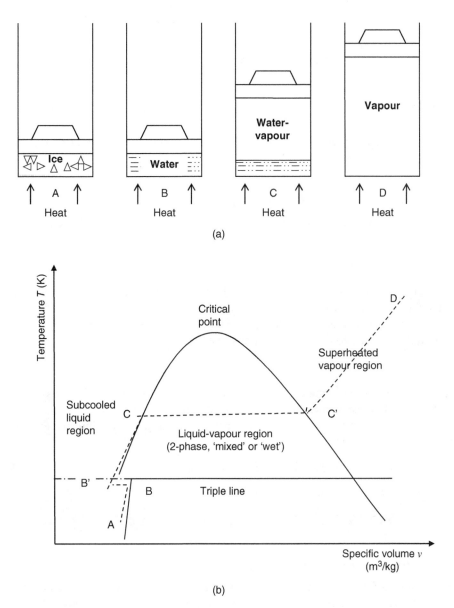

Figure 1.3 (a) Formation of vapour (steam); (b) state diagram; (c) two-phase definitions.

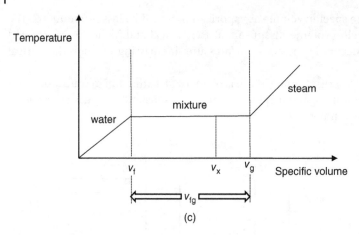

Figure 1.3 *(Continued)*

The term *dryness fraction (x)* is only applicable for the wet region and, as described earlier, is defined as follows:

$$x = \frac{\text{mass of vapour}}{\text{total mass of system (liquid + vapour)}}$$

The total mass of the system = mass of vapour + mass of liquid, hence the system-specific volume along the two-phase line (Figure 1.3c) is:

$$v = (1 - x)v_f + xv_g \tag{1.1}$$

Values of v_f and v_g and other properties for real substances are normally given in tables. Suffix 'f' refers to the liquid; suffix 'g' refers to the dry vapour.

At point C, $x = 0$ and thus $v = v_f$ whereas at point C', $x = 1$ and thus $v = v_g$

Letting suffix 'fg' refer to the change of phase, i.e. $v_{fg} = v_g - v_f$ then

$$v \text{ (at } x) = (1 - x)v_f + xv_g = v_f + x(v_g - v_f) = v_f + xv_{fg} \tag{1.2}$$

Either equation will give the same answer, however the second requires less data manipulation and hence is easier for manual calculation.

Similar expressions can be written for the specific enthalpy, specific internal energy and specific entropy of a substance:

$$h = (1 - x)h_f + xh_g = h_f + x(h_g - h_f) = h_f + xh_{fg}$$

$$u = (1 - x)u_f + xu_g = u_f + x(u_g - u_f) = u_f + xu_{fg}$$

$$s = (1 - x)s_f + xs_g = s_f + x(s_g - s_f) = s_f + xs_{fg}$$

1.5.1 Property Tables and Charts for Vapours

Tables giving data for saturated liquid and saturated vapour conditions, as well as the superheated values of v, u, h and s at a given pressure and over a range of temperatures are readily available. An example is shown in Table 1.1.

Table 1.1 Typical steam table.

T	P = 2.00 MPa (212.42 °C)		
	v	h	s
(°C)	m³/kg	kJ/kg	kJ/kg K
Sat. liquid	0.00117	908.8	2.4474
Sat. vapour	0.09963	2799.5	6.3409
225	0.10377	2835.8	6.4147
250	0.11144	2902.5	6.5453
300	0.12547	3023.5	6.7664
350	0.13857	3137.0	6.9563
400	0.15120	3247.6	7.1271
500	0.17568	3467.6	7.4317

Note in the example used in Table 1.1 that the saturation temperature, i.e. the liquid–vapour phase change temperature, is supplied in brackets after the pressure value.

Some property tables for water/steam are supplied in Appendix A.

Charts for steam that indicate h or T as ordinate against s (specific entropy) as abscissa are also available. An example is shown in Figure 1.4.

1.6 Ideal Gas Behaviour in Closed and Open Systems and Processes

An ideal gas is one which, if pure, will behave according to the ideal gas equation, also known as the equation of state. The equation of state expresses the relationship between pressure (P) and volume (V) as a function of the mass of gas (m), its specific gas constant (R) and the temperature (T) as follows:

$$PV = mRT \tag{1.3}$$

The value of the gas constant depends on the gas molar mass, thus:

$$R = R_o/M \tag{1.4}$$

where R_o is the universal gas constant = 8.314 kJ/kmol K and M is the gas molar mass (kg/kmol).

For gas going through a process between conditions 1 and 2, the equation of state becomes:

$$\frac{P_1 V_1}{T_1} = \frac{P_2 V_2}{T_2} \tag{1.5}$$

An *adiabatic reversible* process is characterized by the following equation: $PV^n = c$

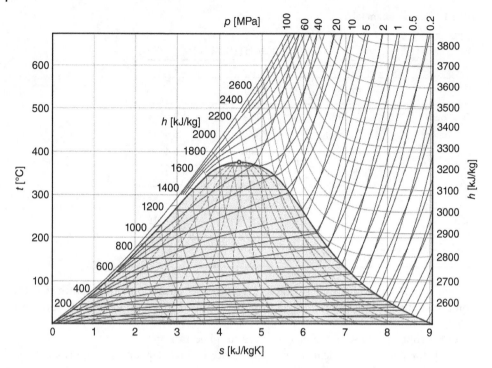

Figure 1.4 Temperature–entropy chart for water/steam. Reproduced with permission from Moran, Shapiro, Boettner and Bailey (2012) *Principles of Engineering Thermodynamics*, 7th edition, John Wiley & Sons.

Therefore, between states 1 and 2, the following relationship can be deduced:

$$P_1 V_1^n = P_2 V_2^n \tag{1.6}$$

where n is the index describing the path of the process.

Combining Equations (1.5) and (1.6), the process can be described in terms of pressure and temperature thus:

$$T_2 = T_1 \left(\frac{P_2}{P_1} \right)^{n-1/n} \tag{1.7}$$

Consider a working fluid undergoing a compression process. Equation (1.7) is used to determine the ideal temperature of the gas at the end of the compression process in terms of the initial temperature and the pressure ratio.

Since entropy is defined as a property that remains constant during an *adiabatic reversible* process, it follows that a temperature–entropy diagram would indicate such a process by a straight line perpendicular to the entropy axis if the process was purely isentropic (Figure 1.5). The friction in an irreversible process will cause the temperature of the gas to be higher than it would have been in a frictionless (reversible) process. Thus, the entropy increases during an *irreversible* process.

The isentropic process (1–2) is found from Equation (1.7) and the relation between the ideal temperature difference and that in the real process (1–2a) forms the isentropic

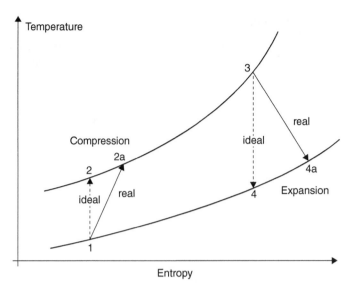

Figure 1.5 Concept of isentropic efficiency.

efficiency for a compressor:

$$\eta_{ic} = \frac{T_2 - T_1}{T_{2a} - T_1} \tag{1.8}$$

Thus, an irreversible compression process requires more work input than an ideal process due to the unwanted temperature rise.

When considering a working fluid undergoing an expansion, a reversible adiabatic or isentropic process (3–4) would, again, be illustrated by a vertical line when plotted on a T–s basis.

A renumbered Equation (1.7) is used to determine the ideal temperature of the gas at the end of the expansion process:

$$T_4 = T_3 \left(\frac{P_4}{P_3} \right)^{n-1/n} \tag{1.9}$$

However, a real expansion (3–4a) would again suffer from irreversibilities, with its final condition having a higher temperature and an increased entropy relative to the isentropic case (see Figure 1.5). In this case, the isentropic efficiency for the expansion is given by:

$$\eta_{it} = \frac{T_3 - T_{4a}}{T_3 - T_4} \tag{1.10}$$

1.7 First Law of Thermodynamics

The first law of thermodynamics is based on the conservation of energy within a system and, between any two state points, can be written as follows:

Initial system energy + Energy in = Final system energy + Energy out (1.11)

1.7.1 First Law of Thermodynamics Applied to Open Systems

The first law applied to an open thermodynamic system, i.e. one where mass may cross a thermodynamic boundary, is termed the Steady Flow Energy Equation (SFEE) and is expressed as:

$$\dot{Q} - \dot{W} = (\Delta H + \Delta KE + \Delta PE) = \dot{m}\left[(h_2 - h_1) + \frac{(\bar{V}_2^2 - \bar{V}_1^2)}{2000} + g\frac{(z_2 - z_1)}{1000}\right]$$

$$(1.12)$$

Here, the LHS represents the heat flow and work transfers. On the RHS, the terms are, respectively: change in enthalpy, kinetic energy change and potential energy change between the start and end states (process).

Note that for an ideal gas, the change in enthalpy can be written as:

$$\Delta h = C_p(T_2 - T_1) \tag{1.13}$$

where C_p is the specific heat capacity at constant pressure (kJ/kg K) for the working fluid.

1.7.2 First Law of Thermodynamics Applied to Closed Systems

The first law applied to a closed thermodynamic system, i.e. one where mass does not cross a thermodynamic boundary, is termed the Non-Flow Energy Equation (NFEE) and is expressed as:

$$Q - W = \Delta U \tag{1.14}$$

where Q and W represent the heat and work during a process and the RHS represents the change in internal energy.

In general, the work done during a non-flow process is represented by the following expression:

$$W = \int PdV \tag{1.15}$$

Note that for ideal gas, the change in internal energy can be written as:

$$\Delta U = mC_v(T_2 - T_1) \tag{1.16}$$

where C_v is the specific heat capacity at constant volume (kJ/kg K) for the gas.

Some important points to note:

- For both SFEE and NFEE there is a sign convention for heat and work. It is common to regard heat leaving a thermodynamic system as heat *loss* (−ve) and heat entering the system as positive. The opposite is used with work; hence, work input to the system is negative whereas work exported is positive.
- In both forms, it is common to regard electrical energy and power as work or work transfer.
- The term *adiabatic* is used to denote zero heat or heat transfer, i.e. Q or $\dot{Q} = 0$.

1.8   Worked Examples

All property values for the following questions can be found in Appendix A.

Worked Example 1.1 – Use of steam tables
Determine the enthalpy of steam at 2 MPa using steam tables for the following conditions:

(a) Dryness fraction $x = 0$.
(b) Dryness fraction $x = 1.0$.
(c) Dryness fraction $x = 0.5$.
(d) At $T = 300\,°C$.

Solution
Given: $P = 2$ MPa.
Find x, T(d).

(a) h (at $x = 0$) = 908.8 kJ/kg
(b) h (at $x = 1$) = 2799.5 kJ/kg
(c) h (at $x = 0.5$) using the two-phase formula

$$h = h_f + x(h_g - h_f)$$

$$= 908.8 + 0.5 \times (2799.5 - 908.8) = 1854.15 \text{ kJ/kg}$$

(d) h (at 300 °C) = 3023.5 kJ/kg

Worked Example 1.2 – Interpolation from steam tables
Determine the enthalpy of steam at 10 MPa and 340 °C.

Solution
Given: $P = 10$ MPa, $T = 340\,°C$.
Find h.

This example illustrates the way in which steam data are estimated when they are not found in the table at the exact conditions.
Let us define the following:

x_1 – initial value (given)
x_2 – final value (given)
y_1 – initial value (given)
y_2 – final value (given)
x_3 – required datum (given)
y_3 – value/property to be found, which can be determined as follows:

$$y_3 = y_1 + (x_3 - x_1) \times \frac{(y_2 - y_1)}{(x_2 - x_1)}$$

Define the parameters for linear interpolation:
$x_1 = 325$; $x_2 = 350$; $y_1 = 2809.1$; $y_2 = 2923.4$; $x_3 = 340$.

Hence,

$$y_3 = y_1 + (x_3 - x_1) \times \frac{(y_2 - y_1)}{(x_2 - x_1)} = 2809.1 + (340 - 325) \times \frac{(2923.4 - 2809.1)}{(350 - 325)}$$

$$= 2877.7 \text{ kJ/kg}.$$

Worked Example 1.3 – Thermodynamic property changes

Steam at 2 MN/m^2 in a saturated dry vapour condition is heated at constant pressure until its temperature reaches a steady value of 400 °C. Determine, for a unit mass (i.e., 1 kg) of the steam:

(a) The change in its volume.
(b) The change in its entropy.
(c) The heat received by the steam.

Solution

Given: $P_1 = 2$ MPa and dry (saturated vapour)
$\qquad P_2 = 2$ MPa, 400 °C.
Find $\Delta v, \Delta s, \Delta h$.
At the beginning of the process (all values refer to the saturated vapour condition):

$v_1 = 0.09963$ m^3/kg
$h_1 = 2799.5$ kJ/kg
$s_1 = 6.3409$ kJ/kg K

At the end state (400 °C, superheated):

$v_2 = 0.15120$ m^3/kg
$h_2 = 3247.6$ kJ/kg
$s_2 = 7.1271$ kJ/kg K

Therefore, the changes during the process are calculated as the difference in values:

(a) $\Delta v = v_2 - v_1 = 0.05157$ m^3/kg (increase)
(b) $\Delta s = s_2 - s_1 = 0.7862$ kJ/kg K (increase)
(c) $\Delta h = h_2 - h_1 = 448.1$ kJ/kg (increase) – this is the heat received.

Worked Example 1.4 – SFEE (steam boiler)

A boiler receives feed water as saturated liquid at 20 bar and delivers steam at 20 bar and 500 °C. If the furnace of this boiler is oil-fired (calorific value of oil = 42 000 kJ/kg), determine the quantity of oil required to process 45 000 kg/h of steam, given a combustion efficiency of 90%.

Solution

Given: $P_1 = P_2 = 20$ bar $= 2$ MPa, $x_1 = 0$, $T_2 = 500$ °C, $CV_{oil} = 4200$ kJ/kg, $\dot{m}_s = 45\ 000$ kg/hr, $\eta_{comb} = 0.9$.
Find $\dot{m}_{oil}$.

This is a constant pressure process, for which the enthalpy values at the initial and final conditions are extracted from a table: $h_1 = h_f = 908.8$ kJ/kg (saturated liquid at 20 bar).

$h_2 = 3467.6$ kJ/kg (superheated steam at 20 bar, 500 °C).

The SFEE for the boiler (no work transfer, and ignoring changes in ΔKE and ΔPE) becomes:

$$\dot{Q} = \dot{m}_s(h_2 - h_1)$$

$$= (45\,000/3600) \times (3467.6 - 908.8) = 32 \text{ MW}$$

The heat generated by burning oil in the furnace is:

Mass of oil burned $\times$ Calorific value $\times$ Efficiency of combustion

Hence, the mass flow rate of oil:

$$\dot{m}_{oil} = \frac{\dot{Q}}{CV_{oil} \times \eta_{comb}} = \frac{32 \times 10^6}{42 \times 10^6 \times 0.9} = 0.846 \text{ kg/s.}$$

Worked Example 1.5 – NFEE (gas compression)

1 kg of air at 1 bar, 27 °C undergoes adiabatic compression to 7 bar. Determine the heat transfer (kJ) during the process.

For air, take $R = 287$ J/kg K, $\bar{C}_v = 714$ J/kg K and $n = 1.4$.

Solution

Given: $m = 1$ kg, $P_1 = 1$ bar, $P_2 = 7$ bar, $R = 287$ J/kg K, $C_v = 714$ J/kg K, $n = 1.4$.
Find Q.

The final temperature is:

$$T_2 = T_1 \left[P_2/P_1 \right]^{\frac{n-1}{n}} = 300 \times \left[7/1 \right]^{\frac{(1.4-1)}{1.4}} = 523 \text{ K}$$

The work done during the process is:

$$W = \int P dV$$

Since $PV^n = C$, $P = CV^{-n}$, i.e.

$$W = C \int V^{-n} dV = \frac{P_2 V_2^n (V_2^{-n+1})}{-n+1} - \frac{P_1 V_1^n (V_1^{-n+1})}{-n+1} = \frac{P_1 V_1 - P_2 V_2}{n-1} = \frac{mR(T_1 - T_2)}{n-1}$$

$$W = \frac{1 \times 287 \times (300 - 523)}{1.4 - 1} = -160\,000 \text{ J}$$

The internal energy change is:

$$\Delta U = m.C_v(T_2 - T_1) = 1 \times 714(523 - 300) = +159287 \text{ J}$$

According to the first law:

$$Q - W = \Delta U$$

Hence, $Q = \Delta U + W = +159287 - 160000 = -780$ J
The low value is not surprising, as the process is adiabatic.

Worked Example 1.6 – SFEE (steam turbine)

A steam turbine receives 120 kg/minute of steam at 1 MPa, 350 °C and a velocity of 20 m/s. The exit condition from the turbine is 100 kPa, $x = 1$ and a velocity of 10 m/s. Neglecting changes in potential energy and assuming the process to be adiabatic, estimate the power output (kW) from the turbine.

Solution

Given: $P_1 = 1$ MPa, $P_2 = 100$ kPa $= 0.1$ MPa, $T_1 = 350$ °C, $\bar{V}_1 = 20$ m/s, $\bar{V}_2 = 10$ m/s, $\dot{m}_s = \frac{120 \text{ kg/min}}{60 \text{ s/min}} = 2$ kg/s.

Assumptions: $\dot{Q} = 0$ (adiabatic), $g(z_2 - z_1) = 0$.

Find $\dot{W}$.

From steam tables:

$h_1 = 3157.7$ kJ/kg
$h_2 = 2675.5$ kJ/kg

Using the SFEE:

$$\dot{Q} - \dot{W} = \dot{m}_s \left[(h_2 - h_1) + \frac{\bar{V}_2^2 - \bar{V}_1^2}{2 \times 1000} + g(z_2 - z_1) \right]$$

$$0 - \dot{W} = \frac{120}{60} \left[(2675.5 - 3157.7) + \frac{10^2 - 20^2}{2 \times 1000} + 0 \right]$$

Hence, $\dot{W} = +964.7$ kW.

Worked Example 1.7 – Isentropic gas expansion

Air is expanded *isentropically* in a nozzle from 13.8 bar and 150 °C to a pressure of 6.9 bar. The process occurs under steady-flow, steady-state conditions. Calculate the exit velocity from the nozzle given that the nozzle is installed in a horizontal plane and that the inlet velocity is 10 m/s. Take the mass flow rate as 1 kg/s, specific heat capacity at constant pressure as 1.005 kJ/kg K and $n = 1.4$.

Solution

Given: $P_1 = 13.8$ bar, $P_2 = 6.9$ bar, $T_1 = 150$ °C, $\bar{V}_1 = 10$ m/s, $\dot{m}_a = 1$ kg/s, $C_p = 1.005$ kJ/kg K, $n = 1.4$.

Assumptions: $\dot{Q} = 0$ (adiabatic), $g(z_2 - z_1) = 0$, $\dot{W} = 0$ (no moving parts).

Find $\bar{V}_2$.

$$T_2 = T_1 \left(\frac{P_2}{P_1} \right)^{n-1/n} = (150 + 273) \times \left(\frac{6.9}{13.8} \right)^{0.4/1.4} = 347 \text{ K}$$

The situation is an open system for which the SFEE applies:

$$\dot{Q} - \dot{W} = \dot{m}_a \left[(h_2 - h_1) + \frac{(\bar{V}_2^2 - \bar{V}_1^2)}{2000} + g \frac{(z_2 - z_1)}{1000} \right]$$

$$= \dot{m}_a \left[C_p(T_2 - T_1) + \left(\frac{\bar{V}_2^2 - \bar{V}_1^2}{2000} \right) + g \left(\frac{z_2 - z_1}{1000} \right) \right]$$

Hence, the SFEE reduces to:

$$0 = \dot{m}_a \left[C_p(T_2 - T_1) + \left(\frac{\bar{V}_2^2 - \bar{V}_1^2}{2000} \right) \right]$$

$$0 = 1 \times \left[1.005(347 - 423) + \frac{1}{2000} \left(\bar{V}_2^2 - 10^2 \right) \right]$$

$$\bar{V}_2 = 390 \text{ m/s.}$$

Worked Example 1.8 – SFEE (gas compression)

A compressor takes in air at 1 bar and 20 °C and discharges into a line. The average air velocity in the line at a point close to the discharge is 7 m/s and the discharge pressure is 3.5 bar. Assuming that the compression occurs adiabatically, calculate the work input (kW) to the compressor. Assume that the air inlet velocity is very small.

For air, take $C_p = 1005$ J/kg K, and $n = 1.4$.

Solution

Given: $P_1 = 1$ bar, $T_1 = 20\,°C = 293$ K, $P_2 = 3.5$ bar, $\bar{V}_2 = 7$ m/s , $\dot{m}_a = 1$ kg/s, $C_p = 1005$ J/kg K, $n = 1.4$.

Assumptions: $\dot{Q} = 0$ (adiabatic), $g(z_2 - z_1) = 0$ (horizontally mounted), $\bar{V}_1 = 0$.

Find $\dot{W}$

For adiabatic compression, the final temperature is calculated as:

$$T_2 = T_1 \left\{ \frac{P_2}{P_1} \right\}^{(n-1)/n} = 293 \times \left(\frac{3.5}{1} \right)^{0.4/1.4} = 419.1 \text{ K}$$

Using the SFEE:

$$\dot{Q} - \dot{W} = \dot{m}_a \left[(h_2 - h_1) + \left(\frac{\bar{V}_2^2 - \bar{V}_1^2}{2000} \right) + g \left(\frac{z_2 - z_1}{1000} \right) \right]$$

$$-\dot{W} = \dot{m}_a \left[C_p(T_2 - T_1) + \left(\frac{\bar{V}_2^2 - \bar{V}_1^2}{2000} \right) \right]$$

Hence,

$$\dot{W} = -1 \times \left[1005 \times (419.1 - 293) + \left(\frac{7^2 - 0}{2000} \right) \right]$$

$$= -[126\ 728 + 0.0245]$$

$$= -126.7 \text{ kW}$$

Worked Example 1.9 – NFEE (work)

An ideal gas occupies a volume of 0.5 m³ at a temperature of 340 K and a given pressure. The gas undergoes a constant pressure process until the temperature decreases to 290 K. Determine:

(a) The final volume.
(b) The work done during the process if the pressure is 120 kPa.

Solution

Given: $V_1 = 0.5 \text{ m}^3$, $T_1 = 340 \text{ K}$, $P = 120 \times 10^3 \text{ Pa}$, $T_2 = 290 \text{ K}$.

Find V_2, W.

(a) For an isobaric process, $P = $ constant so the ideal gas equation for the process is written as:

$$\frac{V_1}{T_1} = \frac{V_2}{T_2}$$

Therefore:

$$V_2 = V_1 \times \frac{T_2}{T_1} = 0.5 \times \frac{290}{340} = 0.426 \text{ m}^3$$

(b) $W = \int P dV$

For a constant pressure process:

$$W = P \int dV = P(V_2 - V_1)$$

$$= 120 \times 10^3 (0.426 - 0.5)$$

$$= -8823 \text{ J} = -8.823 \text{ kJ}.$$

The negative sign indicates that work is imported from an external source to the system.

Worked Example 1.10 – SFEE (mass flow)

Steam at a pressure of 6 MPa and a temperature of 500 °C enters an adiabatic turbine with a velocity of 100 m/s and expands to a pressure of 100 kPa. The steam leaves the turbine as just dry saturated vapour, with a velocity of 10 m/s. If the turbine is required to develop 1 MW, determine the mass flow rate of steam (kg/s) through the turbine.

Solution

Given: $P_1 = 6 \text{ MPa}$, $T_1 = 500 \,°\text{C}$, $P_2 = 100 \text{ kPa} = 0.1 \text{ MPa}$, $\bar{V}_1 = 100 \text{ m/s}$, $\bar{V}_2 = 10 \text{ m/s}$, $\dot{W} = 1 \text{ MW} = 1000 \text{ kW}$.

Assumptions: $\dot{Q} = 0$ (adiabatic), $g(z_2 - z_1) = 0$.

Find $\dot{m}$

From steam tables:

$h_1 = 3422.2 \text{ kJ/kg}$

$h_2 = 2675.5 \text{ kJ/kg}$

$$\dot{Q} - \dot{W} = \dot{m} \left[(h_2 - h_1) + \frac{1}{2} \left(\bar{V}_2^2 - \bar{V}_1^2 \right) + g(z_2 - z_1) \right]$$

Neglecting the changes in potential energies, $\dot{Q} = 0$ for an adiabatic process:

$$-\dot{W} = \dot{m}_s \left[(h_2 - h_1) + \frac{1}{2000} \left(\bar{V}_2^2 - \bar{V}_1^2 \right) \right]$$

$$-1000 = \dot{m}_s \left[(2675.5 - 3422.2) + \frac{10^2 - 100^2}{2 \times 1000} \right]$$

$$\dot{m}_s = 1.33 \text{ kg/s}.$$

1.9 Tutorial Problems

Thermodynamic properties of water and steam are available in Appendix A.

1.1 Determine the specific enthalpy of steam (kJ/kg) at 20 MPa using steam tables for the following conditions:
(a) Dryness fraction $x = 0$.
(b) Dryness fraction $x = 1.0$.
(c) Dryness fraction $x = 0.5$.

[Answers: 1826.3, 2409.7, 2118 kJ/kg]

1.2 Steam at 4 MPa, 400 °C expands at constant entropy until its pressure is 0.1 MPa. Determine, using steam tables, the dryness fraction of the steam at the end of the expansion process and the energy liberated per kilogram of steam under 100% isentropic conditions.

[Answers: 0.902, 758 kJ/kg]

1.3 Repeat Problem 1.2 for when the expansion process is 80% isentropic.

[Answers: 0.969, 606 kJ/kg]

1.4 An ideal gas occupies a volume of 0.5 m³ at 600 K and 100 kPa. The gas undergoes a constant pressure process until its temperature reaches 300 K. Determine:
(a) The final volume (m³).
(b) The work done during the process (kJ).

[Answers: 0.25 m³, −25 kJ]

1.5 *Self-ignition* would occur in an engine using a certain type of gasoline if the temperature due to compression reached 350 °C. Assume inlet conditions of 27 °C and 1 bar. Calculate the highest ratio of compression that may be used to avoid pre-ignition (base your maximum temperature to 1 degree below that of self-ignition) if the law of compression is: $PV^{1.3} = C$.

[Answer: 11.36]

1.6 A gas turbine operates between 9 bar and 1 bar. Helium is the working fluid, expanding adiabatically from an initial temperature of 1000 K and a mass flow of 0.1 kg/s. Determine the exit temperature (K) and the work done (kW) if changes in kinetic energy and potential energy across the turbine are negligible. For Helium, $R = 2.08$ kJ/kg K and $C_p = 5.19$ kJ/kg K.

[Answers: 414 K, 304 kW]

1.7 A turbine manufacturer claims that his turbine will have an output of about 9 MW under the following conditions:
Steam flow 10 kg/s at 6 MPa and 500 °C inlet condition.
100% isentropic expansion to a final pressure of 100 kPa.
The heat transfer from the casing to the surroundings represents 1% of the overall change of enthalpy of the steam.

The exit is 5 m above entry.
The initial velocity of the steam is 100 m/s whereas the exit velocity is 10 m/s.
Do you agree with the manufacturer or not? Support your answer by calculations.

[Answer: yes]

1.8 A compressor takes in air at 1 bar and 20 °C and discharges at an average air velocity of 10 m/s and a pressure of 5 bar. Assuming that the compression occurs isentropically, calculate the specific work input to the compressor (kJ/kg).
Assume that the air inlet velocity is very small. Take, for air, $C_p = 1.005$ kJ/kg K and $n = 1.4$.

[Answer: −172 kJ/kg]

1.9 A steam turbine receives 120 kg/minute of steam at 1 MPa, 350 °C and a velocity of 20 m/s. The exit condition from the turbine is 10 kPa, $x = 1$ and a velocity of 10 m/s. Neglect changes in potential energy and assume the process to be adiabatic. Estimate the power output from the turbine.

[Answer: +1135 kW]

1.10 Find the change in value of specific enthalpy (kJ/kg) for water (steam) between an initial state where $P = 100$ kPa, saturated liquid, and a final state where $P = 100$ kPa and $T = 350$ °C.

[Answer: 2758.85 kJ/kg]

2

Vapour Power Cycles

2.1 Overview

The majority of power stations rely on phase changes in a working fluid such as water to provide power. Thermodynamic analyses such as the Carnot cycle and the Rankine cycle are employed in order to calculate the power output and efficiency of thermal conversion. This chapter will introduce the basic power cycles and examine enhancements that increase overall efficiency.

Learning Outcomes

- To understand the basic vapour cycles and demonstrate the limitation of the hypothetical Carnot cycle.
- To understand the practical amendments used to improve steam cycle efficiency.
- To understand the concept of the Organic Rankine cycle.
- To be able to appraise combined heat and power (CHP), trigeneration and quad generation.
- To be able to solve problems related to steam cycles.

2.2 Steam Power Plants

A power plant is an installation for the production of electrical energy. A steam power plant utilizes the heat released during the combustion of fuel to heat water flowing inside tubes, thus transforming it into steam, which is subsequently allowed to expand in a turbine unit especially designed with rows of blades absorbing the energy of the steam. The mechanical (rotational) energy of the turbine is converted into electrical energy by the generator that is mounted on the same shaft as the turbine unit. A typical steam power plant is shown in Figure 2.1.

After losing its bulk energy, the expanded steam is condensed and returned to the boiler for reuse for two reasons: steam power plants require huge quantities of water and the quality of water required has to be of the highest standard in order to maintain a high operational efficiency of the steam turbine blades and surfaces.

Conventional and Alternative Power Generation: Thermodynamics, Mitigation and Sustainability, First Edition. Neil Packer and Tarik Al-Shemmeri.
© 2018 John Wiley & Sons Ltd. Published 2018 by John Wiley & Sons Ltd.

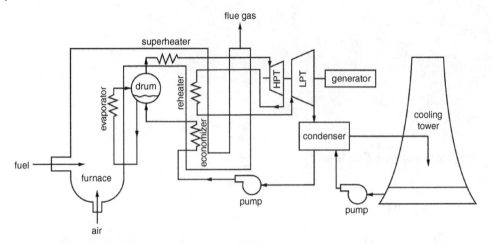

Figure 2.1 Typical steam power plant.

2.3 Vapour Power Cycles

There are five steam power cycles:

a. The Carnot cycle
b. The simple Rankine cycle
c. The Rankine superheat cycle
d. The Rankine reheat cycle
e. The regenerative cycle.

The organic Rankine cycle is a modern addition to the above list. It is based on the standard Rankine cycle with one modification: it employs a different working fluid to water/steam.

These cycles will be analysed from a thermodynamic point of view, and therefore, it is important to define the various parameters used in the process of evaluation.

The following parameters are energy-related terms based on unit mass flow of steam through the cycle and the numbers attached to properties are only relevant to a simple cycle which consists of a pump (1–2), a boiler (2–3), a turbine (3–4) and a condenser (4–1). As will be shown later, a modified and expanded set of notation is required for more sophisticated cycles. However, the following definitions can be extended to accommodate prevailing circumstances:

(a)   Pump work, $w_p = v_{f1} \times \Delta P = (h_2 - h_1)$ (2.1)

(b)   Turbine work, $w_t = (h_3 - h_4)$ (2.2)

(c)   Heat supplied in boiler, $q_{in} = (h_3 - h_2)$ (2.3)

(d)   Heat rejected in condenser, $q_{out} = (h_4 - h_1)$ (2.4)

(e)   Net work, $w_{net} = w_t - w_p$ (2.5)

(f) $\qquad$ Work ratio, $WR = \dfrac{w_t - w_p}{w_t}$ $\hfill$ (2.6)

The term *work ratio* gives an indication of the proportion of the net work to that produced in the turbine. This term ideally needs to be close to unity.

(g) $\qquad$ Cycle efficiency, $\eta_{cycle} = \dfrac{w_t - w_p}{q_{in}} \times 100\%$ $\hfill$ (2.7)

The cycle efficiency is defined as the ratio of net work to heat supplied. In other words, it represents the percentage of fuel energy converted into useful output energy.
   High efficiency is, of course, desirable.

(h) $\qquad$ Specific steam consumption, $SSC = \dfrac{3600}{w_{net}} \left(\dfrac{kg}{kWh} \right)$ $\hfill$ (2.8)

This term indicates the relative size of the plant, i.e. mass flow per unit work output. Of course, the lower the value of this term, the better.

2.3.1   The Carnot Cycle

The Carnot cycle (see Figure 2.2) is a theoretical cycle consisting of four processes:

- The working fluid, initially as two-phase liquid/vapour, is compressed to a high pressure, thus causing it to liquefy, and the liquid water is delivered to the next stage – the boiler.
- The high-pressure saturated liquid water is heated in the boiler until it changes phase to a pure saturated vapour.
- The high-pressure, high-temperature dry saturated water vapour is directed onto the blades of the steam turbine. Hence, the steam energy is converted into mechanical energy driving an electric generator that is coupled on the same shaft as the turbine.

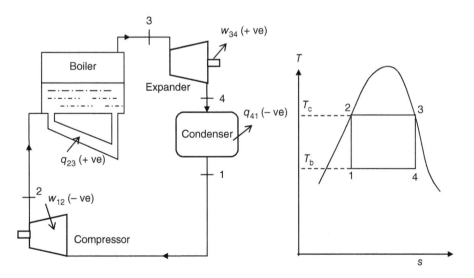

Figure 2.2 The Carnot cycle.

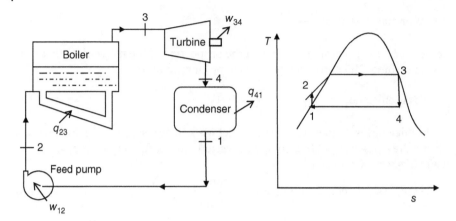

Figure 2.3 The simple Rankine cycle.

- Finally, the exhausted fluid (a two-phase liquid/vapour mixture) is partially condensed in the condenser and then returned to the compressor to complete the cycle.

2.3.2   The Simple Rankine Cycle

The simple Rankine cycle (named in honour of Scottish physicist William John Macquorn Rankine and shown in Figure 2.3) is a practical cycle designed to overcome the shortcomings of the Carnot cycle and consists of four processes:

- The working fluid, initially liquid water, is pumped (with a pressure increase but to a lesser extent than the compressor in the Carnot cycle) and delivered to the next stage – the boiler.
- The high-pressure liquid water is heated in the boiler until it changes phase to water vapour.
- The high-pressure, high-temperature fluid is directed onto the blades of the steam turbine. Hence, the steam energy is converted into mechanical energy driving an electric generator that is coupled on the same shaft as the turbine.
- Finally, the exhausted fluid is completely condensed and returned to the pump to complete the cycle.

Note the two differences between this and the Carnot cycle:

1. The condenser liquefies the water completely, so that, at outlet, the water is pure liquid, which is easier to control and achieve.
2. The compressor in the Carnot cycle is now replaced by a pump dealing with pure water, hence there is a smaller unit dealing with a single-phase fluid.

2.3.3   The Rankine Superheat Cycle

The Rankine superheat cycle (see Figure 2.4) is a practical cycle designed to improve the simple Rankine cycle; it also consists of four processes:

- The working fluid, initially liquid water, is pumped (with an overall pressure increase) to the next unit – the boiler.

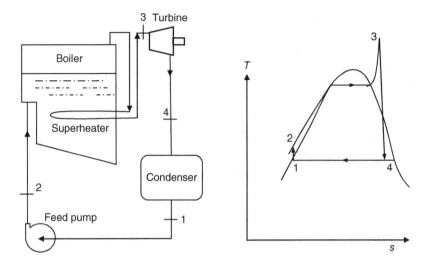

Figure 2.4 The Rankine superheat cycle.

- The high-pressure liquid water is heated in the boiler until it changes phase to water vapour. The vapour is further heated in a *superheater*.
- The high-pressure, high-temperature superheated steam is directed onto the blades of the steam turbine. Hence, the steam energy is converted into mechanical energy driving an electric generator that is coupled on the same shaft as the turbine.
- Finally, the exhausted fluid is completely condensed and returned to the pump for recycling.

Note the difference between this and the simple Rankine cycle: superheating the steam will increase the work output in the turbine, and the quality of steam at turbine exit can be maintained as dry saturated or slightly superheated to avoid damage to the blades of the turbine. However, this is accompanied by an increase in the heat supplied; the pump work is not affected.

2.3.4 The Rankine Reheat Cycle

The Rankine reheat cycle (see Figure 2.5) is a practical cycle designed to improve the Rankine cycle; it consists of the following processes:

- The working fluid, initially liquid water, is pumped (increasing its pressure) and delivered to the next unit – the boiler.
- The high-pressure liquid water is heated in the boiler until it changes phase to water vapour. The vapour is further heated in a superheater.
- The high-pressure, high-temperature superheated steam is allowed to expand across a (high-pressure, HP) turbine to an intermediate pressure above that of the condenser, i.e. an incomplete expansion. The steam is returned to the boiler and is reheated.
- The steam is returned to a separate (low-pressure, LP) turbine and allowed to complete its expansion down to the condenser pressure. The HP and LP turbines are connected to a common shaft.
- Finally, the exhausted fluid is condensed and returned to the pump for recycling.

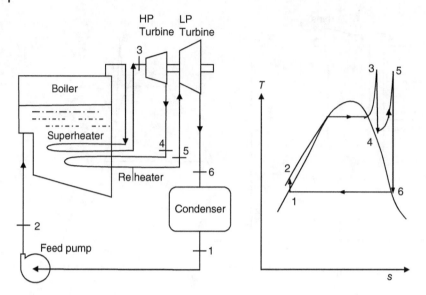

Figure 2.5 The Rankine reheat cycle.

2.3.4.1 Analysis of the Rankine Reheat Cycle

The difference between this and the Rankine superheat cycle is in the fact that multi-stage expansion (two stages or more are common) increases the work output in the turbine and the quality of steam at turbine outlets. However, this is accompanied by an increase in the heat supplied (during the reheat stage). The pump work is not affected. In reference to Figure 2.5, the following are defined:

1. The cycle efficiency

$$\eta_{\text{cycle}} = \frac{w_{34} + w_{56} - w_{12}}{q_{23} + q_{45}} \tag{2.9}$$

Cycle efficiency indicates the extent of conversion of fuel into useful work output, so it is desirable to have this as high as possible.

2. Work ratio

$$WR = \frac{w_{34} + w_{56} - w_{12}}{w_{34} + w_{56}} \tag{2.10}$$

The work ratio indicates the useful fraction of work generated, so it is desirable to have this as close as possible to unity.

3. Specific steam consumption

$$SSC = \frac{3600}{w_{34} + w_{56} - w_{12}} \tag{2.11}$$

This term indicates the compactness of the system, so it is desirable to have this as low as possible. A low value implies less energy is wasted in pumping, heating, etc.

Figure 2.6 Effects of irreversibilities in a steam cycle.

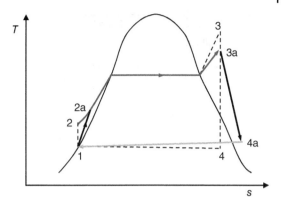

2.3.5 Real Steam Processes

Real steam cycles suffer a significant deviation from the ideal counterparts, especially during the pumping and expansion processes. Losses due to friction and irreversibilities cause an increase in temperature of the end states, and the real processes are, instead, indicated by (1–2a) and (3a–4a), respectively. Pressure drops in the condenser and the boiler are also noticeable but less significant.

The effect of these on a Rankine cycle is shown in Figure 2.6.

Pump – The pump suffers an increase in entropy, defined by the pump's isentropic efficiency (η_{ip}):

$$\eta_{ip} = \frac{h_2 - h_1}{h_{2a} - h_1} \tag{2.12}$$

Turbine – The turbine suffers an increase in entropy, defined by the turbine's isentropic efficiency (η_{it}):

$$\eta_{it} = \frac{h_3 - h_{4a}}{h_3 - h_4} \tag{2.13}$$

Boiler – Only a slight decrease in pressure occurs due to friction losses, so the pressure at exit to the boiler is lower than that at the pump's outlet.

Condenser – An inlet pressure slightly higher than that at the pump's inlet is necessary to overcome friction.

Of the four above effects, the isentropic efficiency of the turbine is by far the most important factor. The drop in pressure in both the boiler and the condenser is often neglected.

2.3.6 Regenerative Cycles

If water were preheated before it entered the boiler or steam generator, less energy would be added in the boiler. If some of the steam were taken from the turbine before it reached the condenser and used to heat the boiler feed-water, two purposes would be accomplished: one, the preheating would occur with no extra energy input, and two, the latent heat of vaporization would not be lost from the system in the condenser. A steam cycle using this type of feed-water heating is called a *regenerative cycle* (see Figure 2.7).

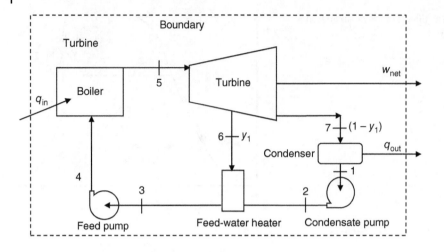

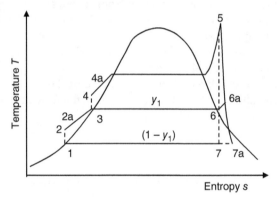

Figure 2.7 Single-feed-heater cycle.

2.3.6.1 Single Feed Heater

With reference to the schematic plant cycle and its T–s diagram presented in Figure 2.7, y_1 is the fraction of the total mass in the system, which leaves the turbine and enters the feed-water heater.

Consideration of energy balance across the boundary of the feed heater gives:

$$y_1 h_6 + (1 - y_1)h_2 = 1 \times h_3$$

Rewriting this gives:

$$y_1 = \frac{h_3 - h_2}{h_6 - h_2} \tag{2.14}$$

The turbine work *per unit mass flow rate* in the system is:

$$w_t = 1 \times (h_5 - h_6) + (1 - y_1)(h_6 - h_7)\,\text{kJ/kg} \tag{2.15}$$

There are two pumps – a condensate pump and a feed pump. Hence, there are two power inputs:

$$w_{p1} = (h_2 - h_1)(1 - y_1)$$

$$w_{p2} = 1 \times (h_4 - h_3)$$

The total pump's work *per unit mass flow rate* in the system is:

$$w_p = w_{p1} + w_{p2} = (h_2 - h_1)(1 - y_1) + 1(h_4 - h_3) \tag{2.16}$$

2.3.6.2 Multiple Feed Heaters

If one feed-water heater is good, then two are better, and so on. This is true, but only to a point. The efficiency does improve as more feed-water heaters are used, but the gain in efficiency is offset by the increase in capital cost and maintenance. Usually, small plants, such as those on ships, have two regenerative heaters, and large, stationary plants have six. Ideally, the feed-water temperature increases in equal increments across each feed heater located between the condenser and the steam generator. The saturated steam temperature in the boiler and the condenser temperature are the two temperature limits used.

Consider a power plant with three feed-water heaters, as shown in Figure 2.8.

Apply an energy balance on the third-stage heater:

$$y_1 h_3 = (1 - y_1)h_{11} = 1 \times h_1, \text{ hence}$$

$$y_1 = \frac{h_1 - h_{11}}{h_3 - h_{11}} \tag{2.17}$$

An energy balance on the second-stage heater yields:

$$y_2 h_4 + (1 - y_1 - y_2)h_9 = (1 - y_1) \times h_{10}, \text{ hence}$$

$$y_2 = \frac{(1 - y_1)(h_{10} - h_9)}{h_4 - h_9} \tag{2.18}$$

Perform an energy balance on the first-stage heater, as follows:

$$y_3 h_5 + (1 - y_1 - y_2 - y_3)h_8 = (1 - y_1 - y_2)h_9, \text{ hence}$$

$$y_3 = \frac{(1 - y_1 - y_2)(h_9 - h_8)}{h_5 - h_8} \tag{2.19}$$

The turbine work per unit mass flow of steam in the system is:

$$w_T = 1 \times (h_2 - h_3) + (1 - y_1)(h_3 - h_4) + (1 - y_1 - y_2)(h_4 - h_5)$$

$$+ (1 - y_1 - y_2 - y_3)(h_5 - h_6)$$

The system pump work is found by adding the work of the condensate and feed pumps.

$$w_{p1} = (h_8 - h_7)(1 - y_1 - y_2 - y_3)$$

$$w_{p2} = (h_{11} - h_{10})(1 - y_1)$$

$$w_p = w_{p1} + w_{p2}$$

$$W_{net} = W_T - W_P$$

The heat supplied per kilogram is given by:

$$q_{in} = h_2 - h_1 \tag{2.20}$$

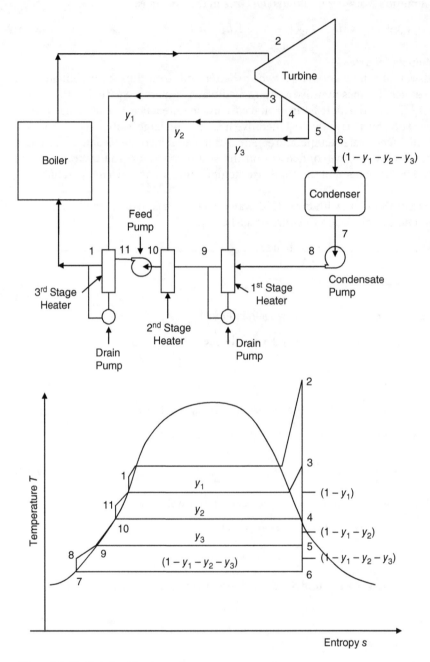

Figure 2.8 Multiple-feed-heater cycle.

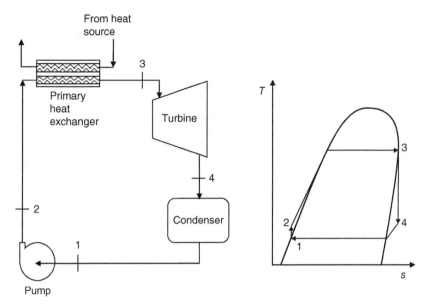

Figure 2.9 Organic Rankine cycle.

The overall thermal efficiency (%) is given by:

$$\eta_{th} = \frac{w_{net}}{q_{in}} \tag{2.21}$$

2.3.7 Organic Rankine Cycle (ORc)

The organic Rankine cycle (see Figure 2.9) is based on the traditional Rankine cycle with one exception. Instead of having water/steam as the working fluid, the ORc system uses an organic fluid, characterized by having a molar mass higher than water, which leads to the possibility of operating with lower thermal input. The advantage of the ORc is realized from the fact that it can operate on waste heat or fuel of lower calorific value than natural gas or coal, such as biomass, solar pond, geothermal source, etc.

2.3.7.1 Choice of the Working Fluid for ORc

The selection of the working fluid is of key importance in low-temperature Rankine cycles. Because of the low temperature, heat transfer inefficiencies are highly prejudicial. In order to recover low-grade heat, the fluid generally has a lower boiling temperature than water. Refrigerants and hydrocarbons are the two most commonly used components. Characteristics to consider are:

- Isentropic saturation vapour curve: Since the purpose of the ORc focuses on the recovery of low-grade heat power, a superheated approach like the traditional Rankine cycle is not appropriate. Therefore, a small amount of superheating at the exhaust of the evaporator (or primary heat exchanger) will always be preferred, which disadvantages 'wet' fluids (that are in a two-phase state at the end of the expansion). In the case of dry fluids, a regenerator should be used.
- Low freezing point, high stability temperature: Unlike water, organic fluids usually suffer chemical deterioration and decomposition at high temperatures. The

maximum hot source temperature is, therefore, limited by the chemical stability of the working fluid. The freezing point should be lower than the lowest temperature in the cycle.

- High heat of vaporization and density: A fluid with a high latent heat and density will absorb more energy from the source in the evaporator and thus reduce the required flow rate, the size of the facility and the pump consumption.
- Low environmental impact: The main parameters taken into account are the ozone depletion potential (ODP) and the global warming potential (GWP).
- Safety: The fluid should be non-corrosive, non-flammable and non-toxic. The ASHRAE safety classification of refrigerants can be used as an indicator of the fluid's level of dangerousness.
- Good availability and low cost.
- Acceptable pressures.

2.4 Combined Heat and Power

Cogeneration, also known as combined heat and power or CHP, is an efficient, cleaner and more responsible approach to generating power and thermal energy from a single fuel source.

Cogeneration uses heat that is otherwise discarded from conventional power generation to produce thermal energy. This 'waste' energy is used to provide cooling or heating for industrial facilities, district energy systems and commercial buildings. By recycling this waste heat, cogeneration systems achieve typical effective electric efficiencies of 50 to 80% – a dramatic improvement over the average 33% efficiency of conventional fossil-fuelled power plants. The higher efficiency of a cogeneration plant reduces air emissions of nitrous oxides, sulphur dioxide, mercury, particulate matter and carbon dioxide, which is the leading greenhouse gas associated with climate change.

Cogeneration technology is not the latest industry buzzword being touted as the solution to our energy woes. Cogeneration is a proven technology that has been around for over 100 years. The first commercial power plant was a cogeneration system designed and built in New York (in 1882) by Thomas Alva Edison. Primary fuels commonly used in cogeneration include natural gas, oil, diesel fuel, propane, coal, wood, wood waste and biomass. These *primary* fuels are used to make electricity, which is a *secondary* fuel. This is why electricity, when compared on an energy cost basis, is typically 3–5 times more expensive than primary fuels such as natural gas.

Cogeneration plants can operate in three scenarios, described below.

2.4.1 Scenario One: Power Only

This is the scenario covered in all the cycles discussed so far, whose sole purpose is to convert a proportion of the input energy transferred to the working fluid into useful work. The remaining portion of the heat is rejected to rivers, lakes, oceans or the atmosphere as waste heat, because its quality (or grade) is too low to be of any practical use. Wasting a large amount of heat is a price paid to produce work, because electrical or mechanical work is the only form of energy on which many engineering devices can operate. Power-only cycles have typical efficiencies in the region of 35%.

A simple power-only plant is shown in Figure 2.10.

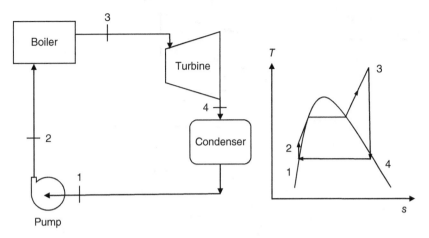

Figure 2.10 A simple power-only plant.

2.4.2 Scenario Two: Heat Only

Many systems or devices, however, require energy input in the form of heat, called *process heat*. Some industries that rely heavily on process heat are chemical, pulp and paper, oil production and refining, steel making, food processing and textiles. Process heat in these industries is usually supplied by steam at 5 to 7 bar and 150 to 200 °C. Energy is usually transferred to the steam by burning coal, oil or natural gas in a furnace.

The generation of high-quality steam for process heating only is generally efficient. However, as in any practical situation, it suffers from losses due to:

Inefficient heat transfer surfaces
Losses in distribution
Irrecoverable thermal energy of the steam at exit from the process heater.
In spite of the above losses, a typical heat-only scheme has an efficiency in excess of 80%. A process-heating-only plant is shown in Figure 2.11.

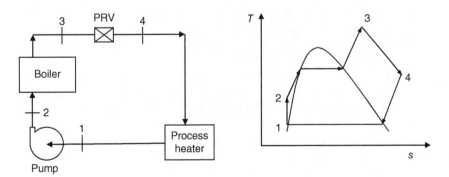

Figure 2.11 A process-heating-only plant.

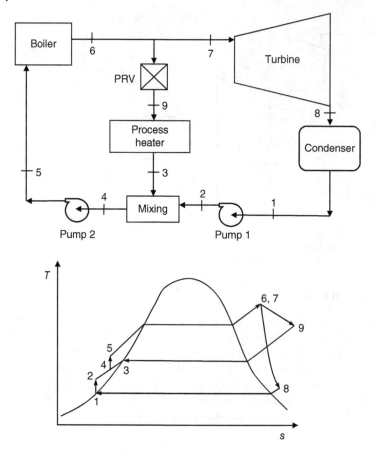

Figure 2.12 Combined heat and power plant.

2.4.3 Scenario Three: Heat and Power

In practice, industries that use large amounts of process heat also consume a large amount of electrical power. Therefore, it makes economical as well as environmental sense to use the already-existing work potential to produce power while utilizing thermal energy for process heat, instead of letting it go to waste. Such a plant is called a *cogeneration plant.* In general, cogeneration is *the production of more than one useful form of energy from the same energy source.* A schematic of a typical steam-turbine cogeneration plant is shown in Figure 2.12.

Under normal operation, some steam is extracted from the turbine at some predetermined intermediate pressure (P_9). The rest of the steam expands to the condenser pressure (P_8) and is then cooled at constant pressure. The heat rejected from the condenser represents the waste heat for the cycle. At times of high demand for process heat, all the steam is routed to the process-heating units and none to the condenser. The steam leaving the boiler is throttled by an expansion or pressure-reducing valve (PRV) to the extraction pressure (P_9) and is directed to the process-heating unit. No power is produced in this mode. When there is no demand for process heat, all the steam passes through the turbine and the condenser, and the cogeneration plant operates as an ordinary steam power plant.

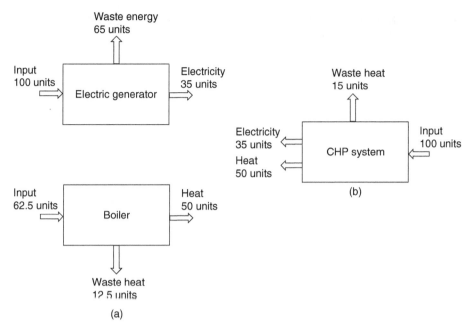

Figure 2.13 Comparison of (a) separate heat and power systems with (b) CHP.

The term *utilization factor* (ε_u) is used to describe the performance of such a plant.

$$\varepsilon_u = \frac{\dot{Q}_{process} + \dot{W}_{net}}{\dot{Q}_{in}} \tag{2.22}$$

2.4.4 Cogeneration, Trigeneration and Quad Generation

Consider the three systems shown in Figure 2.13. Assuming a modern power station to have a thermal efficiency of 35% and an industrial boiler to have an efficiency of 80%, a load of 35 units electrical and 50 units heat would require (35/0.35 = 100) + (50/0.80 = 62.5) = 162.5 units of fuel for separate heat and power generation. In a CHP plant, 100 units of fuel can be used to meet both the 50 units of heat and 35 units of electrical power, giving an overall saving of fuel of 62 units (38%), as illustrated in Figure 2.13.

The opportunity to maximize the outputs from a single energy input is extended to include the use of waste heat as a source of power input to drive an absorption refrigeration system to provide refrigeration and air-conditioning. In addition, quad generation adds the use of carbon dioxide capture for use in the production of sparkling drinks such as cola and lemonade.

2.5 Steam Generation Hardware

Steam power plants operating on the cycles discussed earlier invariably will have many things in common, but due to the diversity of manufacturers in this field, there are differences between plants operating on the same cycle.

The following sections are devoted to discussing the different designs for the main components of a typical power station.

2.5.1   Steam Boiler Components

The boiler unit:

- Steam drum/or water separator
- Header
- Riser
- Down-comer
- Feed-water pump
- Steam circulation pump (with forced circulation systems).

The furnace unit:

- Fuel preparation and feeder
- Air fan
- Air–fuel mixer
- Firing system (burner).

Heat transfer units:

- Evaporator
- Superheater
- Reheater
- Economizer
- Air preheater.

Boiler accessories

Boiler accessories are needed for adequate control and operation, such as:

- Water-level gauges
- Safety valves
- Soot blowers
- Feed-water regulators
- Boiler blow down valves
- CCTV: closed-circuit system for remote viewing of water-level gauges, smoke discharges from stacks and flame conditions in furnaces.

Some of the heat exchange components are described in more detail below.

The *boiler/furnace* comprises the space needed for combustion and heat transfer to the water/steam mixture. A boiler/furnace also collects a portion of any ash at the furnace bottom and allows for its removal. The furnace wall is made of fire-resistant bricks and is protected by *water-filled tubes*. These tubes constitute the major part of heat-absorbing surfaces in the furnace.

The *superheater* is a heat exchanger designed to increase the temperature of the saturated steam leaving the furnace. Superheaters may be radiant, convective or combined, depending on the mode of heat transfer. A convective superheater is normally screened from the furnace radiation by a bank of water-filled tubes. Radiant superheaters are usually placed on the furnace wall.

Superheating may be split into two stages for reasons of space optimization. A radiant superheater may be used as a secondary superheater.

The *reheater* is essentially a convective type and is usually located in the space between the primary and secondary superheaters. The function of a reheater is to increase the temperature of partially expanded steam by returning it to the boiler. The temperature of the reheated steam is usually equal to the superheated steam temperature.

The *economizer* is a heat exchanger unit that enables the heating of feed-water by the exhaust gases before leaving the furnace and therefore increases the thermal efficiency of the plant.

The *air preheater* is a heat exchanger used to preheat the air for combustion using exhaust gases and hence improve the firing process and combustion of the system.

2.5.2 Types of Boiler

The function of a boiler is to convert fuel into thermal energy as efficiently as possible. According to design, boilers are classified into two groups (see Figure 2.14):

- Fire-tube boilers (mass transport applications – locomotives)
- Water-tube boilers (power station boilers).

All power station boilers are of the water-tube type. Water circulates within the tubes and partially becomes steam. There are two arrangements for water-tube boilers:

- *Natural circulation boilers:* Water from the boiler drum flows downward to the bottom of the evaporator tubes.
- *Forced circulation boilers:* A pump is used to increase the flow rate of water from the drum and supply it to the headers at the bottom of the boiler.

2.5.3 Fuel Preparation System

Power plants may use solid, liquid or gaseous fuels and, as such, there has to be some provision for fuel storage, handling and preparation.

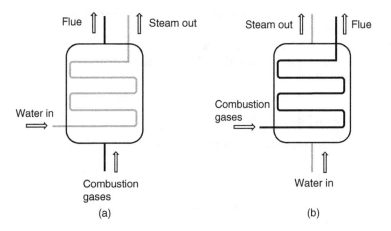

Figure 2.14 Types of boiler. (a) Water-tube boiler; (b) fire-tube boiler.

- In coal-fired power stations, the raw feed coal from the coal storage area is first crushed into small pieces and then conveyed to the coal feed hoppers at the boilers. The coal is next pulverized into a very fine powder and is injected into the combustion zone assisted by a fan.
- Some power stations burn fuel oil rather than coal. The oil must be kept warm (above its pour point) in the fuel oil storage tanks to prevent it from congealing and becoming unpumpable. The oil is usually heated to about 100 °C before being pumped through the furnace fuel oil spray nozzles.
- Boilers in many power stations use natural gas as their main fuel. Other power stations may use processed natural gas as auxiliary fuel in the event that their main fuel supply (coal or oil) is interrupted. In such cases, separate gas burners are provided on the boiler furnaces.

2.5.4 Methods of Superheat Control

Temperature regulation and control are important for both superheaters and reheaters. Steam temperature adjustments are frequently made at the time of boiler commissioning. The principal methods are an addition or reduction of heat transfer surfaces. Steam temperature can also be adjusted by regulating the hot gas temperature and mass flow rate. These are generally accomplished by changing the excess air or the effectiveness of the evaporation section.

The most effective approaches (see Figure 2.15) are:

- *Gas bypass with a single burner:* This method is employed for regulating convection superheaters. At higher loads, the bypass is opened, allowing some of the gas to flow around by opening the damper.
- *Auxiliary burners (dual burners):* An additional burner (or burners) is used when a higher degree of superheat is required.
- *Tilting burner:* The burner can be tilted up or down to be directed to or away from the superheaters. This will result in a change in the rate of heat absorption and, hence, the gas temperature.
- *Desuperheater:* If the degree of superheat is too high, it can be reduced in a desuperheater by using a spray of water on the coil of the superheated steam so that the degree of superheat is reduced. If it becomes wet (two-phase), it can be further superheated by secondary superheaters.
- *Gas recirculation:* By using a fan, the flue gas is returned to the furnace in such a way that it releases its heat only to the superheaters. This will provide an increase in the steam outlet temperature.

2.5.5 Performance of Steam Boilers

Three criteria are used to define/specify the performance of industrial boilers: boiler efficiency, the boiler rating and equivalent evaporation.

2.5.5.1 Boiler Efficiency
The sole function of the boiler is to transfer energy from fuel to water and convert it into high-pressure steam. Boilers are designed to do this job as efficiently as possible. The evaluation of boiler efficiency is derived as follows:

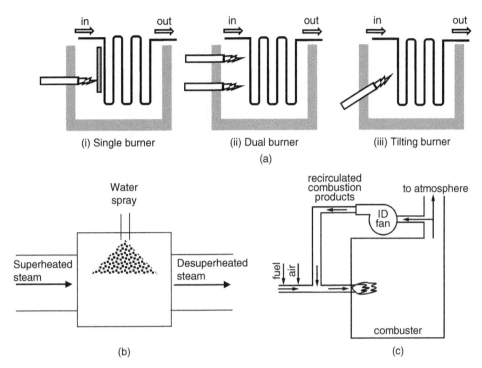

Figure 2.15 Superheat control techniques. (a) Burner arrangements; (b) desuperheater; (c) gas recirculation.

Rate of energy input:

$$\dot{Q}_{in} = \dot{m}_{fuel} \times CV_{fuel} \times \eta_{comb} \tag{2.23}$$

Rate of energy output:

$$\dot{Q}_{out} = \dot{m}_s(h_s - h_{feed-water}) \tag{2.24}$$

Boiler efficiency:

$$\eta_b = \frac{\dot{Q}_{out}}{\dot{Q}_{in}} \tag{2.25}$$

Fuel burners and furnaces are designed to achieve almost complete combustion of fuel. Modern water-tube boilers have a large heat transfer area and are well insulated from the surroundings, therefore combustion efficiencies as high as 90% are possible.

2.5.5.2 Boiler Rating

The capacity of the boiler is stated as the maximum amount of steam generated in kg/hour. Boiler capacity depends on the design of the boiler firing system. Important factors are the size of the combustion zone, the type of heat exchangers used and their material, the type of fuel used and its feed rate as well as the air-to-fuel ratio.

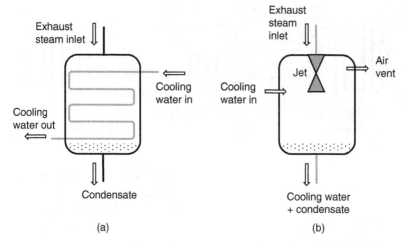

Figure 2.16 Types of steam condenser. (a) Non-contact condenser; (b) evaporative jet condenser.

2.5.5.3 Equivalent Evaporation

Equivalent evaporation is defined as the ratio of energy transferred to the water (Q_{out}) to the latent heat of evaporation (h_{fg}) at 100 kPa and 100 °C, which is 2256.7 kJ/kg.

Thus, the equivalent evaporation (EE) is given by:

$$EE = \frac{Q_{out}}{2256.7} \tag{2.26}$$

2.5.6 Steam Condensers

The function of the condenser is to receive the exhaust steam from the turbine, condense it and deliver the condensate to the feed pump so that it can be reused.

There are two types of condenser (see Figure 2.16), the non-contact condenser and the jet condenser, but there are a number of variations within the general classification.

A (shell and tube type) surface condenser uses cooling water, from a cooling tower or river, passing through the condenser tubes, with the steam condensing on the outside of the tubes.

With jet condensers, the cooling water is mixed with the steam, and the two streams may be arranged to meet at the top of the condenser, or the water may enter at the top and steam enter at the bottom. The counter current type is the more effective option. The mixing of steam and cooling water may be a disadvantage if the water is untreated and is not of sufficiently good quality for boiler feed-water.

2.5.6.1 Condenser Calculations

The condenser is a constant-pressure system and energy transfers will therefore appear as changes in enthalpy. Thus, neglecting unaccounted losses and applying conservation of energy across the boundary:

Energy in = Energy out

$$\dot{m}_s h_{s1} + \dot{m}_{cw} h_{cw1} = \dot{m}_s h_{s2} + \dot{m}_{cw} h_{cw2} \tag{2.27}$$

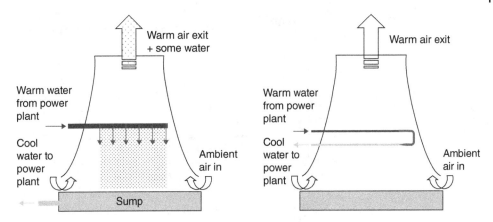

Figure 2.17 Cooling towers.

Hence, the amount of cooling water required is calculated by rewriting Equation 2.27 as:

$$\dot{m}_{cw} = \frac{m_s(h_{s1} - h_{s2})}{C_{p,cw} \times (T_{cw2} - T_{cw1})}$$ (2.28)

2.5.7 Cooling Towers

Steam power stations require large quantities of cooling water in order to condense the steam after its passage through the steam turbine so that the water can be reused. One of the most effective methods is to use the principle of evaporative cooling in a cooling tower (see Figure 2.17). The warm water is sprayed into the tower near the top. It is then allowed to fall through a packing of wooden slats, which break up the stream and provide a large wetted surface to facilitate evaporation. A current of air passes up the tower, either naturally as in a chimney, or induced by a fan. The warm water is cooled, mainly by evaporation, while the air is raised in temperature and saturated, or nearly saturated, with water vapour. Alternatively, in dry towers, the steam is circulated inside a heat exchanger around the inside of the tower and is cooled indirectly by the rising air. Dry towers are a convenient method used necessarily in hotter climates or in circumstances where cheap water is not available in large quantities.

2.5.8 Power-station Pumps

2.5.8.1 Pump Applications

Power plants need pumps for various services with varying specifications, generally defined by the flow rate, working head and efficiency. These services require a variety of designs using different principles of operation and structural details. Power-plant pumps may be classified into three categories (see Figure 2.18):

- Reciprocating
- Centrifugal
- Rotary.

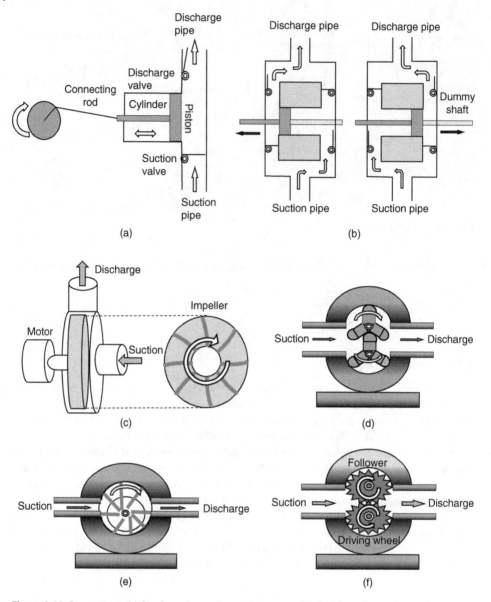

Figure 2.18 Pump types. (a) Single-acting reciprocating pump; (b) double-acting reciprocating pump; (c) centrifugal pump; (d) lobe pump; (e) vane pump; (f) gear pump.

a. Reciprocating Pumps This type of pump operates by using a reciprocating piston or diaphragm. The liquid enters a pumping chamber via an inlet valve and is pushed out via an outlet valve by the action of the piston or diaphragm.

Reciprocating pumps are generally very efficient and are suitable for very high heads at low flows. The reciprocating pump is not tolerant of solid particles and delivers a highly pulsed flow. The discharge flow system has to include an accumulator to provide even flows. Piston pumps can be single-acting or double-acting (see Figures 2.18a and 2.18b).

b. Centrifugal Pumps Centrifugal pumps are the most common type used in water distribution, with medium head and medium flow rates. Centrifugal pumps use an aerofoil bladed structure called the *impeller* (see Figure 2.18c) that is mounted on a centrally supporting shaft, driven externally by a motor.

c. Rotary Pumps These pumps utilize the simple geometry of lobes, vanes or toothed gears (see Figures 2.18d, 2.18e and 2.18f) to trap the liquid and transfer it from inlet to outlet.

2.5.9   Steam Turbines

The energy in high-pressure steam from a boiler or other source spins a steam turbine, and the turbine spins the shaft of a generator. All 'steam-electric' generating plants use this basic process whether powered by coal, natural gas, oil, nuclear or geothermal energy.

The principle of a steam turbine is to allow steam to flow through a nozzle to a lower pressure, converting part of the steam energy to kinetic jet energy. Two turbine configurations are available (see Figure 2.19):

- *Impulse turbine:* Expands the steam in stationary turbine nozzles, where it attains high velocity. The steam jet flows over moving blades or 'buckets' without further expansion. The blades are fastened to the rims of rotating discs mounted on the turbine shaft. The turbine uses velocity and pressure stages.
- *Reaction turbine:* The steam expands in both fixed blades and moving blades on the rim of a rotating drum.

Turbines can be installed in tandem (series) or cross (parallel) arrangements.

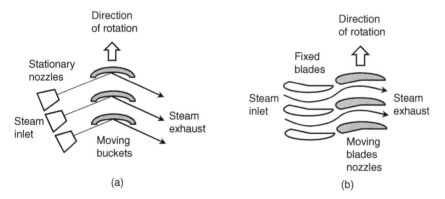

Figure 2.19 Steam turbines. (a) Impulse turbine; (b) reaction turbine.

2.6   Worked Examples

All property values for the following questions can be found in Appendix A.

Worked Example 2.1 – Ideal Carnot cycle
A steam power plant operating on the ideal Carnot cycle has the following data:

- Boiler pressure: 6 MPa
- Condenser pressure: 10 kPa.

Determine the following for unit mass flow rate of steam (1 kg/s):

(a) Input work.
(b) Gross and net work output.
(c) Heat supplied in the boiler.
(d) Heat rejected in the condenser.
(e) Thermal efficiency of the plant.
(f) Work ratio.
(g) Specific steam consumption.

Solution
Given: $P_2 = P_3 = 6$ MPa, $P_1 = P_4 = 10$ kPa $= 0.01$ MPa (see Figure 2.2).
Find w_{12}, w_{34}, q_{23}, q_{41}, η_{cycle} WR, SSC.

At 10 kPa

$$h_f = 191.83 \text{ kJ/kg} \qquad s_f = 0.6493 \text{ kJ/kg K}$$

$$h_g = 2584.70 \text{ kJ/kg} \qquad s_g = 8.1502 \text{ kJ/kg K}$$

At 6 MPa

$$h_f = 1213.35 \text{ kJ/kg} \qquad s_f = 3.0267 \text{ kJ/kg K}$$

$$h_g = 2784.30 \text{ kJ/kg} \qquad s_g = 5.8892 \text{ kJ/kg K}$$

Since $s_1 = s_2$

$$x_1 = \frac{s_2 - s_f}{s_g - s_f} = \frac{3.0267 - 0.6493}{8.1502 - 0.6493} = 0.317$$

Using $h = h_f + x(h_g - h_f)$

$$h_1 = 191.83 + 0.317(2584.7 - 191.83) = 950.25 \text{ kJ/kg}$$

Since $s_4 = s_3$

$$x_4 = \frac{s_3 - s_f}{s_g - s_f} = \frac{5.8892 - 0.6493}{8.1502 - 0.6493} = 0.698$$

$$h_4 = 191.83 + 0.698(2584.7 - 191.83) = 1863.24 \text{ kJ/kg}$$

(a) $w_{12} = h_2 - h_1 = 1213.35 - 950.25 = 263.10 \text{ kJ/kg}$
(b) $w_{34} = h_3 - h_4 = 2784.30 - 1863.42 = 920.88 \text{ kJ/kg}$
  Hence, net work output $= 920.88 - 263.10 = 657.78 \text{ kJ/kg}$.
(c) $q_{23} = h_3 - h_2 = 2784.30 - 1213.35 = 1570.95 \text{ kJ/kg}$
(d) $q_{41} = h_4 - h_1 = 1863.40 - 950.25 = 913.15 \text{ kJ/kg}$
(e) $\eta_{cycle} = \dfrac{w_{net}}{q_{in}} = \dfrac{920.88 - 263.10}{1570.95} \times 100 = 41.87\%$
(f) $WR = \dfrac{w_{net}}{w_t} = \dfrac{920.88 - 263.10}{920.88} = 0.714$
(g) $SSC = \dfrac{3600}{w_{net}} = \dfrac{3600}{920.88 - 263.10} = 5.473 \dfrac{kg}{kWh}$

Worked Example 2.2 – Rankine cycle
A steam power plant operating on the ideal simple Rankine cycle has the following data:

- Boiler pressure: 6 MPa
- Condenser pressure: 10 kPa.

Determine the following for unit mass flow rate of steam (1 kg/s):

(a) Input work.
(b) Gross and net work output.
(c) Heat supplied in the boiler.
(d) Heat rejected in the condenser.
(e) Thermal efficiency of the plant.
(f) Work ratio.
(g) Specific steam consumption.

Solution
Given: $P_2 = P_3 = 6$ MPa, $P_1 = P_4 = 10$ kPa $= 0.01$ MPa (see Figure 2.3).
Find w_{12}, w_{34}, q_{23}, q_{41}, η_{cycle} WR, SSC.
    At 10 kPa

$$h_1 = 191.83 \text{ kJ/kg}$$

$$h_2 = h_1 + v_{f1}(p_2 - p_1)$$
$$= 191.83 + 0.0010102(6 \times 10^3 - 10)$$
$$= 191.83 + 6.05 = 197.88 \text{ kJ/kg}$$

As in Worked Example 2.1, $h_3 = 2784.30$ kJ/kg and $h_4 = 1863.42$ kJ/kg

(a) $w_{12} = h_2 - h_1 = 197.88 - 191.83 = 6.05 \text{ kJ/kg}$
(b) $w_{34} = h_3 - h_4 = 2784.30 - 1863.42 = 920.88 \text{ kJ/kg}$
    Hence, net work output $= 920.88 - 6.05 = 914.83$ kJ/kg.
(c) $q_{23} = h_3 - h_2 = 2784.30 - 197.88 = 2586.51 \text{ kJ/kg}$
(d) $q_{41} = h_4 - h_1 = 1863.42 - 191.83 = 1671.59 \text{ kJ/kg}$
(e) $\eta_{\text{cycle}} = \dfrac{w_{\text{net}}}{q_{\text{in}}} = \dfrac{920.88 - 6.05}{2586.51} \times 100 = 35.36\%$
(f) $WR = \dfrac{w_{\text{net}}}{w_t} = \dfrac{920.88 - 6.05}{920.88} = 0.9935$
(g) $SSC = \dfrac{3600}{w_{\text{net}}} = \dfrac{3600}{920.88 - 6.05} = 3.935 \text{ kg/kWh}$

Worked Example 2.3 – Rankine superheat (isentropic)
A steam power plant operating on the ideal Rankine superheat cycle has the following data:

- Boiler pressure: 6 MPa
- Condenser pressure: 10 kPa
- Boiler exit temperature: 500 °C.

Determine the following for unit mass flow rate of steam (1 kg/s):

(a) Input work.
(b) Gross and net work output.
(c) Heat supplied in the boiler.
(d) Heat rejected in the condenser.
(e) Thermal efficiency of the plant.
(f) Work ratio.
(g) Specific steam consumption.

Solution

Given: $P_2 = P_3 = 6$ MPa, $P_1 = P_4 = 10$ kPa $= 0.01$ MPa, $T_3 = 500\,°$C (see Figure 2.4). Find $w_{12}, w_{34}, q_{23}, q_{41}, \eta_{cycle}$ WR, SSC.

At 10 kPa

$$h_1 = 191.83 \text{ kJ/kg}$$

$$h_2 = h_1 + v_{f1}(p_2 - p_1)$$
$$= 191.83 + 0.0010102(6 \times 10^3 - 10)$$
$$= 191.83 + 6.05 = 197.88 \text{ kJ/kg}$$

At 6 MPa, 500 °C

$$h_3 = 3422.2 \text{ kJ/kg}$$

$$s_3 = 6.8803 \text{ kJ/kg K}$$

At 10 kPa

$$s_f = 0.6493 \text{ kJ/kg K}, \ s_g = 8.1502 \text{ kJ/kg K},$$
$$h_f = 191.83 \text{ kJ/kg}, \ h_g = 2584.7 \text{ kJ/kg}$$

Since $s_4 = s_3$

$$x_4 = \frac{s_4 - s_f}{s_g - s_f} = \frac{6.8803 - 0.6493}{8.1502 - 0.6493} = 0.8307$$

$$h_4 = 191.83 + 0.8307(2584.7 - 191.83) = 2179.59 \text{kJ/kg}$$

(a) $w_{12} = h_2 - h_1 = 197.88 - 191.83 = 6.05 \text{kJ/kg}$
(b) $w_{34} = h_3 - h_4 = 3422.2 - 2179.59 = 1242.61 \text{ kJ/kg}$

   Hence, net work output $= 1242.61 - 6.05 = 1236.56$ kJ/kg.

(c) $q_{23} = h_3 - h_2 = 3422.2 - 197.88 = 3224.41 \text{ kJ/kg}$
(d) $q_{41} = h_4 - h_1 = 2179.59 - 191.83 = 1987.76 \text{kJ/kg}$
(e) $\eta_{cycle} = \dfrac{w_{net}}{q_{in}} = \dfrac{1242.61 - 6.05}{3224.41} \times 100 = 38.34\%$
(f) $WR = \dfrac{w_{net}}{w_t} = \dfrac{1242.61 - 6.05}{1242.61} = 0.995$
(g) $SSC = \dfrac{3600}{w_{net}} = \dfrac{3600}{1242.61 - 6.05} = 2.911 \text{ kg/kWh}$

Worked Example 2.4 – Rankine reheat

A steam power plant operating on the ideal Rankine reheat cycle has the following data:

- Boiler pressure: 6 MPa
- Boiler exit temperature: 500 °C
- Reheat pressure: 200 kPa
- Reheat temperature: 500 °C
- Condenser pressure: 10 kPa.

Determine the following for unit mass flow rate of steam (1 kg/s):

(a) Input work.
(b) Gross and net work output.
(c) Heat supplied in the boiler.
(d) Heat rejected in the condenser.
(e) Thermal efficiency of the plant.
(f) Work ratio.
(g) Specific steam consumption.

Solution

Given: $P_2 = P_3 = 6$ MPa, $P_4 = P_5 = 200$ kPa $= 0.2$ MPa, $P_1 = P_6 = 10$ kPa $= 0.01$ MPa, $T_3 = T_5 = 500$ °C (see Figure 2.5).

Find w_p, w_t, q_{in}, q_{out}, η_{cycle} WR, SSC.

At 10 kPa

$$h_1 = 191.8 \text{ kJ/kg}$$

$$h_2 = h_1 + v_{f1}(p_2 - p_1)$$

$$= 191.83 + 0.0010102(6 \times 10^3 - 10)$$

$$= 191.83 + 6.05 = 197.88 \text{ kJ/kg}$$

At 6 MPa, 500 °C

$$h_3 = 3422.2 \text{ kJ/kg}; \ s_3 = 6.8803 \text{ kJ/kg K}$$

Point 4 is at 200 kPa

$$s_f = 1.5300 \text{ kJ/kg K}, \quad s_g = 7.1272 \text{ kJ/kg K}$$

$$h_f = 504.7 \text{ kJ/kg}, \quad h_g = 2706.7 \text{ kJ/kg}$$

and has the same entropy as Point 3, i.e. $s_4 = s_3 = 6.8803$ kJ/kg K, hence:

$$x_4 = \frac{s_4 - s_f}{s_g - s_f} = \frac{6.8803 - 1.5300}{7.1272 - 1.5300} = 0.956$$

$$h_4 = 504.7 + 0.956(2706.7 - 504.7) = 2609.6 \text{ kJ/kg}$$

At 200 kPa, 500 °C

$$h_5 = 3487.1 \text{ kJ/kg}$$

$$s_5 = 8.5133 \text{ kJ/kg K}$$

Since $s_6 = s_5 > s_g$ at 10 kPa, i.e. superheated.
From tables, $h_6 = 2718.5$ kJ/kg (by interpolation).

(a) $w_p = h_2 - h_1 = 197.88 - 191.83 = 6.05$ kJ/kg
(b) $w_t = (h_3 - h_4) + (h_5 - h_6) = (3422.2 - 2609.6) + (3487.1 - 2718.5) = 1581.2$ kJ/kg

Hence, net work output $= 1581.2 - 6.05 = 1575.15$ kJ/kg.

(c) $q_{in} = (h_3 - h_2) + (h_5 - h_4) = (3422.2 - 197.88) + (3487.1 - 2609.6) = 4101.9$ kJ/kg
(d) $q_{out} = h_6 - h_1 = 2718.5 - 191.83 = 2526.7$ kJ/kg
(e) $\eta_{cycle} = \dfrac{w_{net}}{q_{in}} = \dfrac{1575 - 6.05}{4101.9} \times 100 = 38.4\%$
(f) $WR = \dfrac{w_{net}}{w_t} = \dfrac{1575 - 6.05}{1575} = 0.996$
(g) $SSC = \dfrac{3600}{w_{net}} = \dfrac{3600}{1575 - 6.05} = 2.285$ kg/kWh

Worked Example 2.5 – Rankine superheat (with irreversibilities)

A steam power plant operating on the real Rankine superheat cycle has the following data:

- Boiler pressure: 6 MPa
- Condenser pressure: 10 kPa
- Boiler exit temperature: 500 °C
- Isentropic efficiency of the turbine: 80%
- Isentropic efficiency of the pump: 80%.

Determine the following for unit mass flow rate of steam:

(a) Input work.
(b) Gross and net work output.
(c) Heat supplied in the boiler.
(d) Heat rejected in the condenser.
(e) Thermal efficiency of the plant.
(f) Work ratio.
(g) Specific steam consumption.

Solution

See Figure 2.6.

At 10 kPa
$h_1 = 191.83$ kJ/kg, $h_2 = 197.88$ kJ/kg
Since $\eta_{ip} = (h_2 - h_1)/(h_{2a} - h_1)$
Then

$$h_{2a} = h_1 + (h_2 - h_1)/\eta_{ip} = 191.83 + \frac{197.88 - 191.83}{0.8} = 199.39 \text{ kJ/kg}$$

At 6 MPa, 500 °C

$$h_3 = 3422.2 \text{ kJ/kg}, \quad s_3 = 6.8803 \text{ kJ/kg K}$$

At 10 kPa

$$h_f = 191.83 \text{ kJ/kg}, \quad h_g = 2584.7 \text{ kJ/kg}$$
$$s_f = 0.6493 \text{ kJ/kg}, \quad s_g = 8.1502 \text{ kJ/kg K}$$

Since $s_4 = s_3$

$$x_4 = \frac{s_3 - s_f}{s_g - s_f} = \frac{6.8803 - 0.6493}{8.1502 - 0.6493} = 0.8307$$

$$h_4 = h_f + x_4 h_{fg} = 191.83 + 0.8307 \,(2584.7 - 191.83) = 2179.59 \text{ kJ/kg}$$

$$h_{4a} = h_3 - \eta_{it}(h_3 - h_4) = 3422.2 - 0.8\,(3422.2 - 2179.59) = 2428.1 \text{ kJ/kg}$$

(a) $w_{12} = h_2 - h_1 = 199.39 - 191.83 = 7.56 \text{ kJ/kg}$

(b) $w_{34} = h_3 - h_4 = 3422.2 - 2428.10 = 994.10 \text{ kJ/kg}$

Hence, net work output = $994.10 - 7.56 = 986.54$ kJ/kg.

(c) $q_{in} = q_{23} = h_3 - h_2 = 3422.2 - 199.39 = 3222.81 \text{ kJ/kg}$

(d) $q_{41} = h_4 - h_1 = 2428.10 - 191.83 = 2236.27 \text{ kJ/kg}$

(e) $\eta_{cycle} = \dfrac{w_{net}}{q_{in}} = \dfrac{994.10 - 7.56}{3222.81} \times 100 = 30.6\%$

(f) $WR = \dfrac{w_{net}}{w_t} = \dfrac{994.10 - 7.56}{994.10} = 0.992$

(g) $SSC = \dfrac{3600}{w_{net}} = \dfrac{3600}{994.10 - 7.56} = 3.649 \text{ kg/kWh}$

Worked Example 2.6 – Single stage feed heater (with irreversibilities)

Calculate the cycle efficiency and the net power output for a steam power plant that operates on the Rankine superheat cycle with one feed heater. The plant has a turbine inlet steam condition of 6 MPa and 500 °C; part of the steam is bled to the feed-water heater at 800 kPa and the remainder continues to expand to the condenser at 10 kPa. The turbine has an isentropic efficiency of 90% and the pump is 100% isentropic. The flow rate of steam in the boiler is 50.0 kg/s.

Compare this to a power plant that does not have regenerative feed-water heating.

Solution

Given: $P_4 = P_5 = 6 \text{ MPa}, \; T_5 = 500°\text{C}, \; P_2 = P_6 = 0.8 \text{ MPa}, \; P_1 = P_7 = 0.01 \text{ MPa},$
$\eta_T = 0.9, \; \eta_P = 1.0, \; \dot{M} = 50 \text{ kg/s}.$
Find $w_{net}, \eta_{cycle}.$
See Figure 2.7.
From steam tables:

$$h_5 = 3422.2 \text{ kJ/kg}$$
$$h_1 = 191.83 \text{ kJ/kg}$$
$$h_3 = 721.11 \text{ kJ/kg}$$

Calculate the following:

$x_7 = (6.8803 - 0.6493)/7.5009 = 0.8307$

$h_7 = 191.83 + 0.8307 \times 2392.87 = 2179.58$ kJ/kg

$x_6 = (6.8803 - 2.0462)/4.6166 > 1$, i.e. superheated

$h_6 = 2870$ kJ/kg by interpolation.

At the high-pressure turbine:

$$h_{6a} = h_5 - 0.9(h_5 - h_6) = 3422.2 - 0.9(3422 - 2870) = 2925.22 \text{ kJ/kg}$$

At the low-pressure turbine:

$$h_{7a} = h_5 - 0.9(h_5 - h_7) = 3422.2 - 0.9(3422 - 2179.58) = 2303.84 \text{ kJ/kg}$$

$$h_2 = h_1 + v_{f1}(P_2 - P_1)$$

$$= 191.83 + 0.001014(800 - 10)$$

$$= 192.63 \text{ kJ/kg}$$

$$h_4 = h_3 + v_{f3}(P_4 - P_3)$$

$$= 721.11 + 0.001115(6 \times 10^3 - 800) = 726.91 \text{ kJ/kg}$$

For the regeneration cycle:
Apply first-law analysis to the heater to find y_1.

$$y_1 = \frac{h_3 - h_2}{h_6 - h_2} = \frac{721.11 - 192.63}{2925.22 - 192.63} = 0.193$$

$$w_p = (1 - y_1)(h_2 - h_1) + 1(h_4 - h_3) = 0.65 + 5.8 = 6.45 \text{ kJ/kg}$$

$$w_t = 1(h_5 - h_{6a}) + (1 - y_1)(h_{6a} - h_{7a}) = 496.98 + 507.67 = 1004.65 \text{ kJ/kg}$$

Power output:

$$\dot{W}_{net} = \dot{m}_s(w_T - w_p) = 50(1004.65 - 6.45) = 49.587 \text{ MW}$$

Heat supplied:

$$\dot{Q}_{in} = \dot{m}_s(h_5 - h_4) = 50(3422.20 - 726.87) = 134.766 \text{ MW}$$

Cycle efficiency:

$$\eta_{cycle} = \frac{\dot{W}_{net}}{\dot{Q}_{in}} = \frac{49.587}{134.766} \times 100 = 36.8\%$$

For the non-regenerative cycle:

$$w_t = h_5 - h_7 = 3422.2 - 2303.86 = 1118.36 \text{ kJ/kg}$$

$$w_p = v_{f1}(P_5 - P_1) = 0.001014(6000 - 10) = 6.07 \text{ kJ/kg}$$

$$w_{net} = w_t - w_p = 1118.36 - 6.07 = 1112.29 \text{ kJ/kg}$$

$$h_4 = h_1 + w_p = 191.83 + 6.07 = 197.9 \text{ kJ/kg}$$

$$q_{in} = h_5 - h_4 = 3422.2 - 197.9 = 3224.3 \text{ kJ/kg}$$

$$\eta_{cycle} = \frac{W_{net}}{q_{in}} = \frac{1112.29}{3224.30} \times 100 = 34\%$$

Therefore, the use of the heater improved the basic cycle efficiency by 3%.

Worked Example 2.7 – Multi-stage feed heaters (isentropic)

A steam power plant runs on a three-stage regenerative cycle. The steam enters the turbine at 6 MPa and 500 °C and fractions are withdrawn at 2.0 MPa, 600 kPa and 100 kPa for feed-water heating. The remaining steam exhausts at 10 kPa to the condenser. Compute the overall thermal efficiency of the power plant.

Solution

Given: $P_2 = 6\,\text{MPa}$, $T_2 = 500\,°\text{C}$, $P_3 = 2\,\text{MPa}$, $P_4 = 600\,\text{kPa} = 0.6\,\text{MPa}$, $P_5 = 100\,\text{kPa} = 0.1\,\text{MPa}$, $P_6 = 10\,\text{kPa} = 0.01\,\text{MPa}$.

Find η_{th}.

See Figure 2.8.

Calculate the enthalpies for the states using steam property tables or charts.

$$h_2 = 3422.2\ \text{kJ/kg}; s_2 = 6.8803\ \text{kJ/kg K}$$

$$h_3 = 3100.0\ \text{kJ/kg}$$

$$h_4 = 2610.0\ \text{kJ/kg}$$

$$h_5 = 2497.0\ \text{kJ/kg}$$

$$h_6 = 2180.0\ \text{kJ/kg}$$

$$h_7 = 191.8\ \text{kJ/kg}$$

$$h_9 = 417.46\ \text{kJ/kg}$$

$$h_{10} = 670.56\ \text{kJ/kg}$$

$$h_1 = 908.79\ \text{kJ/kg}$$

$$h_8 = h_7 + v_7 \Delta p = 191.8 + 0.001\,(100 - 10) = 191.9\ \text{kJ/kg}$$

$$h_{11} = h_{10} + v_{10} \Delta p = 670.56 + 0.0011\,(2000 - 600) = 672.1\ \text{kJ/kg}.$$

$$y_1 = \frac{h_1 - h_{11}}{h_3 - h_{11}} = \frac{908.79 - 672.1}{3100.0 - 672.1} = 0.097$$

$$y_2 = \frac{(1 - y_1)(h_{10} - h_9)}{h_4 - h_9} = \frac{(1 - 0.097)(670.56 - 417.46)}{2610.0 - 417.46} = 0.104$$

$$y_3 = \frac{(1 - y_1 - y_2)(h_9 - h_8)}{h_5 - h_8} = \frac{(1 - 0.097 - 0.104)(417.46 - 191.9)}{5 + 5} = 0.078$$

$$w_t = (h_2 - h_3) + (1 - y_1)(h_3 - h_4) + (1 - y_1 - y_2)(h_4 - h_5)$$
$$+ (1 - y_1 - y_2 - y_3)(h_5 - h_6)$$

$$= 1083.6\ \text{kJ/kg}$$

$$w_{p1} = (h_8 - h_7)(1 - y_1 - y_2) = 0.08\,\text{kJ/kg}$$

$$w_{p2} = (h_{11} - h_{10})(1) = 1.5\,\text{kJ/kg}$$

$$w_p = w_{p1} + w_{p2} = 1.58\,\text{kJ/kg}$$

$$w_{net} = \sum w = 1031.0\,\text{kJ/kg}$$

$$q_{in} = h_2 - h_1 = 2513.4\,\text{kJ/kg}$$

$$\eta_{th} = \frac{w_{net}}{q_{in}} = \frac{1031.0}{2513.4} \times 100 = 41\%$$

In the absence of all feed heaters:

$$w_p = v_7(P_2 - P_6) = 0.001(6000 - 10) = 6\ \text{kJ/kg}$$

$$w_t = h_2 - h_6 = 3422.2 - 2180.0 = 1242.2\ \text{kJ/kg}$$

$$w_{net} = w_t - w_p = 1236.2\ \text{kJ/kg}$$

$$q_{in} = h_2 - h_7 = 3422.2 - 191.7 = 3230.5\ \text{kJ/kg}$$

$$\eta_{th} = \frac{w_{net}}{q_{in}} = \frac{1236.2}{3230.5} \times 100 = 38.26\%$$

That is, feed heaters contributed to an increase of 3% net or a relative increase of 9%.

Worked Example 2.8 – CHP (isentropic with throttling)

In a CHP plant, steam enters the turbine at 7 MPa and 500 °C. At turbine exit, the steam has expanded to 10 kPa. Steam is then condensed at constant pressure and pumped to the boiler pressure of 7 MPa. At times of high demand for process heat, some steam leaving the boiler is throttled to 500 kPa and is routed to the process heater. The extraction fractions are adjusted so that steam leaves the process heater as a saturated liquid at 500 kPa. It is subsequently pumped to 7 MPa. The mass flow rate of steam through the boiler is 10 kg/s. Disregarding any pressure drops and heat losses in the piping, and assuming the turbine and the pump to be 100% isentropic, determine:

(a) The rate at which process heat can be supplied when the turbine is bypassed.
(b) The power produced when the process heater is bypassed.
(c) The rate of process heat supply, power output and utilization factor when the PRV is closed, 50% of the steam is extracted from the turbine at 500 kPa for process heating and the remaining 50% expands through to the condenser.

Solution
See Figure 2.12.

$h_1 = 191.83$ kJ/kg (h_f at 10 kPa)
$h_2 = h_1 + v_{f1}(P_2 - P_1) = 191.83 + 0.00101(500 - 10) = 192.3$ kJ/kg
$h_3 = 640.23$ kJ/kg (h_f at 500 kPa)

$$h_4 = \frac{m_2 h_2 + m_3 h_3}{m_2 + m_3} = \frac{5 \times 192.3 + 5 \times 640.12}{5 + 5} = 416.3 \, \text{kJ/kg}$$

$h_5 = h_4 + v_{f3}(P_5 - P_4) = 416.3 + 0.00109 \, (7000 - 500) = 423.3 \, \text{kJ/kg}$

$h_6 = 3410.3 \, \text{kJ/kg}$

$h_7 = 2738.2 \, \text{kJ/kg} \, (x_7 = 0.995)$

$h_8 = 2153 \, \text{kJ/kg} \, (x_8 = 0.819)$

(a) The maximum rate of process heat is achieved when all the steam leaving the boiler is throttled and sent to the process heater and none is sent to the turbine; that is:

$$\dot{Q}_{p,\text{max}} = m_6(h_6 - h_3) = 10(3410.3 - 640.23) = 27.7 \, \text{MW}$$

(b) When no process heat is supplied, all the steam leaving the boiler will pass through the turbine and will expand to the condenser pressure of 10 kPa.

$$\dot{W}_t = \dot{m}_6(h_6 - h_8) = (10) \times (3410.3 - 2153) = 12.571 \, \text{MW}$$

$$\dot{W}_p = \dot{m}_6(h_2 - h_1) = (10) \times (192.32 - 191.83) = 0.0049 \, \text{MW}$$

$$\dot{W}_{\text{net}} = \dot{W}_t - \dot{W}_p = 12.566 \, \text{MW}$$

(c)

  i. Process heat:

$$\dot{Q}_{\text{process}} = \dot{m}_7(h_7 - h_3) = 5(2738.6 - 640.23) = 10.490 \, \text{MW}$$

  ii. Power:

$$\dot{W}_p = \dot{m}_8(h_2 - h_1) + \dot{m}_7(h_5 - h_4)$$
$$= 5(192.32 - 191.83) + 10(423.35 - 416.26) = 0.073 \, \text{MW}$$

$$\dot{W}_t = \dot{m}_6(h_6 - h_7) + \dot{m}_8(h_7 - h_8)$$
$$= 10(3410.3 - 2738.6) + 5(2738.6 - 2153) = 9.644 \, \text{MW}$$

  Hence, $W_{\text{net}} = 9.644 - 0.073 = 9.570 \, \text{MW}$.

  iii. The input heat supplied and utilization factor:

$$\dot{Q}_{\text{in}} = \dot{m}_6(h_6 - h_5)$$
$$= 10(3410.3 - 423.35) = 29.869 \, \text{MW}$$

$$\varepsilon_u = \frac{\dot{Q}_{\text{process}} + \dot{W}_{\text{net}}}{\dot{Q}_{\text{in}}} = \frac{10.490 + 9.570}{29.869} \times 100 = 67\%$$

Worked Example 2.9 – CHP (no throttling with irreversibilities)

A combined heat and power plant provides a small textile factory with 10 MW of process steam and 10 MW of electricity. The boiler generates steam at 10 MPa and 450 °C. Steam expands initially to 500 kPa; at this stage, some steam is extracted from the turbine for process heating. The remaining steam continues to expand to the condenser at 50 kPa. The condensate is pumped to a pressure of 500 kPa and is then mixed with the used process steam (saturated liquid); the mixture, which can be assumed to be saturated liquid at the intermediate pressure, is pumped back to the boiler. Disregarding any pressure drops and heat losses in the piping, and assuming the turbine and the pump to be 90% isentropic, determine:

(a) The mass flow rate of process steam.
(b) The mass flow rate of steam through the boiler.
(c) The utilization factor.

Solution

Given: $\dot{Q}_{process} = 10 \times 10^6$ W, $\dot{W}_{elec} = 10 \times 10^6$ W, $P_6 = 10 \times 10^6$ Pa, $T_6 = 450°C$, $P_2 = P_3 = P_7 = 500 \times 10^3$ Pa, $P_1 = P_8 = 50 \times 10^3$ Pa, $\eta_p = \eta_t = 0.9$ (see Figure 2.12).
Find $\dot{m}_{process}$, $\dot{m}_{boiler}$, η_u.

$h_1 = h_f$ at 50 kPa = 340.5 kJ/kg
$v_1 = v_f$ at 50 kPa = 0.00103 m³/kg
$w_{p1} = v_1(P_2 - P_1)/\eta_p = 0.00103(500 - 50)/0.9 = 0.49$ kJ/kg
$h_2 = h_1 + w_p = 340.5 + 0.49 = 341.0$ kJ/kg
$h_3 = h_f$ at 0.5 MPa = 640.23 kJ/kg
$h_4 = h_3 = 640.23$ kJ/kg (given in the question)
$w_{p2} = v_4(P_5 - P_4)/\eta_p = 0.00109(10\,000 - 500)/0.9 = 11.5$ kJ/kg
$h_5 = h_4 + w_{p2} = 640.23 + 11.5 = 651.7$ kJ/kg
$P_6 = 10$ MPa, $h_6 = 3240.9$ kJ/kg
$T_6 = 450 °C$, $s_6 = 6.419$ kJ/kg K

Point 7 at 0.5 MPa, $s_f = 1.8607$, $s_g = 6.8213$ kJ/kg K, and $s_7 = s_6$, hence:

$$x_7 = \frac{s_7 - s_f}{s_{fg}} = \frac{6.4190 - 1.8607}{6.8213 - 1.8607} = 0.919$$

$h_7 = h_f + x_7 h_{fg} = 640.2 + 0.919(2748.7 - 640.2) = 2577.7$ kJ/kg

$h_{7a} = h_6 - \eta_t(h_6 - h_7) = 3240.9 - 0.9(3240.9 - 2577.7) = 2644.0$ kJ/kg

Point 8 at 0.05 MPa, $s_f = 1.0910$, $s_g = 7.5939$ kJ/kg K, and $s_8 = s_6$, hence:

$$x_8 = \frac{s_8 - s_f}{s_{fg}} = \frac{6.419 - 1.0910}{7.5939 - 1.0910} = 0.819$$

$h_8 = h_f + x_8 h_{fg} = 340 + 0.819(2645.9 - 340) = 2229.4$ kJ/kg

$h_{8a} = h_6 - \eta_t(h_6 - h_8) = 3240.9 - 0.9(3240.9 - 2229.4) = 2330.5$ kJ/kg

(a) Mass flow rate of process steam:

$$\dot{Q}_{process} = \dot{m}_3(h_6 - h_3)$$

$$\dot{m}_3 = \frac{\dot{Q}_{process}}{h_6 - h_3} = \frac{10\,000}{3240.9 - 640.2} = 3.845 \text{ kg/s}$$

(b) Mass flow rate through the boiler:
From turbine considerations

$$\dot{W}_T = \dot{m}_6(h_{6a} - h_{7a}) + (\dot{m}_6 - \dot{m}_3)(h_{7a} - h_{8a})$$

$$10\,000 = \dot{m}_6(3240.9 - 2644) + [(\dot{m}_6 - 3.845)(2644 - 2330.5)]$$

hence

$$\dot{m}_6 = 12.308 \text{ kg/s}$$

(c) Utilization factor:

$$\dot{Q}_{in} = \dot{m}_6(h_6 - h_5) = 12.308(3240.9 - 651.7) = 31.867 \text{ MW}$$

$$\varepsilon_u = \frac{\dot{Q}_{process} + \dot{W}_{net}}{\dot{Q}_{in}} = \frac{10 + 10}{31.867} \times 100 = 62.7\%$$

Worked Example 2.10 – Boiler analysis

A steam boiler supplies 12 000 kg of steam per hour at 5 MPa and 600 °C from feed-water at 5 MPa, saturated liquid. The fuel consumption is 1200 kg of furnace oil per hour. The calorific value of the fuel is 40 000 kJ/kg and the combustion efficiency is 90%. Determine:

(a) The boiler efficiency (%).
(b) The equivalent evaporation (kg/hr).
(c) The equivalent evaporation per kg of fuel used (kg/kg).

Solution

Given:

$$P_2 = 5 \times 10^6 \text{ Pa} = P_1, \ T_2 = 600°\text{C}, \ \dot{m}_{fuel} = 1200 \text{ kg/hr},$$
$$CV_{fuel} = 40 \times 10^3 \text{kJ/kg}, \ \eta_{comb} = 0.9.$$

Find η_{boiler}, *EE, SEE.*

From the steam tables, the enthalpy of steam produced at 5 MPa and 600 °C is:

$$h_s = 3666.5 \text{ kJ/kg}$$

The enthalpy of feed-water is:

$$h_w = h_f \text{ at } 5 \text{ MPa} = 1154.2 \text{ kJ/kg}$$

Energy input:

$$\dot{Q}_{in} = \dot{m}_{fuel} \times CV_{fuel} \times \eta_{comb} = 1200 \times 40000 \times 0.9 = 43.2 \times 10^6 \text{ kJ/hr}$$

Energy output:

$$\dot{Q}_{out} = \dot{m}_s(h_s - h_w) = 12000(3666.5 - 1154.2) = 30.147 \times 10^6 \text{ kJ/hr}$$

(a) Efficiency $= \dfrac{\dot{Q}_{out}}{\dot{Q}_{in}} \times 100 = \dfrac{30.147 \times 10^6}{43.200 \times 10^6} \times 100 = 69\%$

(b) Equivalent evaporation $= \dfrac{\dot{Q}_{out}}{2256.7} = \dfrac{30.147 \times 10^6}{2256.7} = 13359 \text{ kg/hr}$

(c) Specific equivalent evaporation $= \dfrac{\text{Equivalent evaporation}}{1200}$

$$= \frac{13359}{1200} = 11.1 \text{ kg steam per kg fuel.}$$

Worked Example 2.11 – Condenser analysis

A surface condenser operating at a pressure of 50 kN/m^2 condenses 2 tonnes of steam per hour. The steam enters the condenser with a dryness fraction of 0.95 and is condensed, but not undercooled. Cooling water enters the condenser at a temperature of $10\,°\text{C}$ and leaves at $40\,°\text{C}$. Determine the flow rate (tonnes/hr) of the cooling water.

Take the specific heat capacity of water to be 4180 J/kg K.

Solution
Given:

$$P_{s1} = 50\,\text{kPa} = 0.05\,\text{MPa} = P_{s2}(\text{sat liq}), \ x_{s1} = 0.95, \ \dot{m}_s = 2\,\text{tonnes/hr}$$
$$T_{w1} = 10\,°\text{C}, \ T_{w2} = 40\,°\text{C}, \ C_p = 4180\,\text{J/kg K}.$$

Find $\dot{m}_{cw}$.

Steam enthalpies from tables:

$$h_{s1} = 340.5 + 0.95(2645.9 - 340.5) = 2530.6 \text{ kJ/kg}$$
$$h_{s2} = 340.5 \text{ kJ/kg}$$

Change in cooling water enthalpy:

$$h_{w1} - h_{w2} = C_p \Delta T = 4.18(40 - 10) = 125.4 \text{ kJ/kg}$$

Hence, the mass flow of water can be calculated by equating the heat transferred from the steam and the heat gained by the water:

$$\dot{m}_{cw} = \frac{\dot{m}_s(h_{s1} - h_{s2})}{h_{cw2} - h_{cw1}} = \frac{2(2530.6 - 340.5)}{125.4} = 34.93 \ \frac{\text{tonnes}}{\text{hr}}.$$

2.7 Tutorial Problems

2.1 A hypothetical steam plant operates on the Carnot cycle between pressures of 3 MPa and 10 kPa. Determine:
  (a) The thermal efficiency.
  (b) The work ratio.
  (c) The specific steam consumption.

  [Answers: 37%, 0.787, 5.4 kg/kWh]

2.2 The steam plant of Problem 2.1 is modified to operate on the simple Rankine cycle between the same pressure limits. Determine:
  (a) The thermal efficiency.
  (b) The work ratio.
  (c) The specific steam consumption.

  [Answers: 32.3%, 0.9964, 4.271]

2.3 The steam plant of Problem 2.2 is modified such that a superheater is used to raise the temperature of steam to $400\,°\text{C}$. Determine, for this plant:
  (a) The thermal efficiency.
  (b) The work ratio
  (c) The specific steam consumption.

  [Answers: 34.1%, 0.9973, 3.477]

2.4 The steam plant of Problem 2.3 is modified to operate on the reheat cycle under the same lower and upper conditions, with reheat pressure and temperature of 200 kPa and 400 °C, respectively. Assume all processes are 100% isentropic. Determine, for this plant:
(a) The thermal efficiency.
(b) The work ratio.
(c) The specific steam consumption.

[Answers: 34.4%, 0.998, 2.840kg/kWh]

2.5 A steam power plant operating on the real Rankine superheat cycle has the following data:
- Boiler pressure: 3 MPa
- Condenser pressure: 10 kPa
- Boiler exit temperature: 400 °C
- Isentropic efficiency of the turbine: 80%
- Isentropic efficiency of the pump: 80%.

Determine:
(a) The thermal efficiency of the plant.
(b) The work ratio.
(c) The specific steam consumption.

[Answers: 27.24%, 0.9954, 4.354]

2.6 A two-stage turbine receives steam at 5 MPa and 360 °C. The steam expands to 1.4 MPa in the high-pressure turbine, then is reheated at 1.4 MPa to 300 °C and finally is expanded through the low-pressure turbine to a condenser pressure of 10 kPa. The isentropic efficiency of the two turbine stages is 85% and that for the pump is 90%. If the power output is 1.0 MW, determine:
(a) The mass flow rate of steam through the boiler.
(b) The boiler capacity.

[Answers: 0.965 kg/s, 3.172 MW]

2.7 A steam power plant operates on the Rankine superheat cycle with one feed heater. The plant has a turbine inlet steam condition of 6 MPa and 500 °C. Part of the steam is bled to the feed-water heater at 800 kPa and the remainder continues to expand to the condenser at 10 kPa. Both the turbine and the pump have an isentropic efficiency of 100%. The flow rate of steam in the boiler is 10.0 kg/s.
(a) Calculate the cycle efficiency and the net power output for this system.
(b) Compare this to a power plant that does not have regenerative feed-water heating.

[Answers: (a) 40.8%, 11 MW; (b) 38.4%, 12.36 MW]

2.8 A steam power plant operates on the Rankine reheat cycle with one feed heater. The plant has a turbine inlet steam condition of 6 MPa and 500 °C. Part of the steam is bled to the feed-water heater at 800 kPa and the remainder is reheated at this pressure until its temperature reaches 500 °C, and then expands to the condenser at 10 kPa. Both turbine stages and the pump have an isentropic efficiency of 80%.

(a) Determine the ideal mixing ratio of regenerative steam and the corresponding cycle efficiency.

(b) A company sales representative proposes that this plant would have a better efficiency if the ratio of the feed heater's mass flow rate was fixed to 0.4. Do you accept the sales representative's advice or not? Demonstrate by calculating the new efficiency.

[Answers: 0.1893, 47.9%, don't agree, 40.5%]

2.9 In a CHP plant, steam enters the turbine at 7 MPa and 500 °C. Some steam is extracted from the turbine at 500 kPa for process heating. The remaining steam continues to expand to 10 kPa. Steam is then condensed at constant pressure and pumped to the boiler pressure of 7 MPa. At times of high demand for process heat, some steam leaving the boiler is throttled to 500 kPa and is routed to the process heater. The extraction fractions are adjusted so that steam leaves the process heater as a saturated liquid at 500 kPa. It is subsequently pumped to 7 MPa. The mass flow rate of steam through the boiler is 10 kg/s. Disregarding any pressure drops and heat losses in the piping, and assuming the turbine and the pump to be 90% isentropic, determine:

(a) The rate at which process heat can be supplied when the turbine is bypassed.

(b) The power produced when the process heater is bypassed.

(c) The rate of process heat supply and power output when the PRV is closed, 50% of the steam is extracted from the turbine at 500 kPa for process heating and the remaining 50% expands through to the condenser. Determine also the utilization factor for this case.

[Answers: 27.7 MW, 11.3 MW, 10.8 MW, 8.6 MW, 0.65]

2.10 A salesman claims that he can supply you with a condenser that will require less than 15 tonnes of water for every tonne of steam under the following conditions:
  • Surface condenser operating at a pressure of 50 kPa.
  • The steam enters the condenser with a dryness fraction of 1.0 and is condensed to a saturated liquid ($x = 0$).
  • Cooling water enters the condenser at a temperature of 10 °C and leaves at 50 °C. Assume $C_{pw} = 4.18$ kJ/kg K.

Do you agree or disagree? Support your answer by calculations.

[Answer: Agree, as the rate of water is 13.79 tonnes for every tonne of steam]

3

Gas Power Cycles

3.1 Overview

Gas power cycles use air as the working fluid; there are two main categories of gas power cycles, namely, gas turbines and gas engines. Both internal combustion and external combustion types are discussed in this chapter.

Enhancement to power cycles will be reviewed and calculations will demonstrate the increase in efficiency of some key changes to the basic cycles.

Learning Outcomes

- To understand the basic gas turbine cycle processes.
- To understand the practical amendments employed to improve the basic gas turbine cycle efficiency, for example, intercooling, reheating and regeneration.
- To be able to appraise the efficiency of a gas turbine power plant.
- To be familiar with Otto, Diesel, dual combustion and Stirling engine cycles.
- To be able to solve problems related to gas power cycles.

3.2 Introduction to Gas Turbines

Gas turbines utilize the principles of gas compression, heat addition and gas expansion to generate power. Gas turbines used in smaller energy utilities can provide a fast response and high efficiency coupled with low emissions.

3.3 Gas Turbine Cycle

The closed-cycle ideal gas turbine plant is shown in Figure 3.1. The figure also shows the temperature–entropy diagram of the thermodynamic cycle, comprising four processes:

Process 1–2: An ideal compression (isentropic process, i.e. $s_2 = s_1$).
Process 2–3: The fuel is burnt with the air at constant pressure.

Conventional and Alternative Power Generation: Thermodynamics, Mitigation and Sustainability,
First Edition. Neil Packer and Tarik Al-Shemmeri.
© 2018 John Wiley & Sons Ltd. Published 2018 by John Wiley & Sons Ltd.

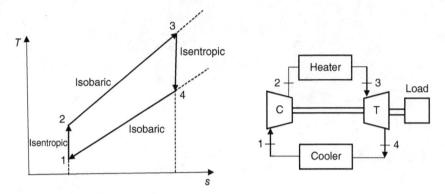

Figure 3.1 Ideal gas turbine cycle.

Process 3–4: The high-energy gas is allowed to expand against the turbine blades, producing useful work. This process ideally runs at constant entropy (i.e. $s_4 = s_3$). Process 4–1: The working gas is cooled back to T_1 at constant pressure.

3.3.1 Irreversibilities in Gas Turbine Processes

Thermodynamic realities, for example, friction and heat transfer, introduce some deviation from the ideal isentropic paths taken by the compression and the expansion processes discussed previously. There is also a pressure drop during heat addition, due to friction between the combustion gases and the surfaces. The pressure at the end of expansion is slightly higher than atmospheric to ensure the free flow of the system. Figure 3.2 highlights these real processes.

3.3.2 The Compressor Unit

If there is little change in velocity and if the compressor is mounted horizontally then both kinetic energy and potential energy terms are negligible. Assuming the compression is adiabatic, the Steady Flow Energy Equation (conservation of energy) will reduce to:

$$-\dot{W}_c = \dot{m}_a(h_2 - h_1) \tag{3.1}$$

As before, the frictionless, adiabatic (i.e. isentropic) temperature at the end of the compression is calculated using:

$$T_2 = T_1\left(\frac{P_2}{P_1}\right)^{(n-1)/n} \tag{3.2}$$

(For an isentropic process, the value of n is given by the ratio of the specific heats, i.e. C_p/C_v, and in some texts is represented by the symbol γ).

The connection between the frictionless, adiabatic compression temperature, T_2, and the real compression temperature, T_{2a}, is indicated by rearranging the isentropic efficiency equation (Equation (1.8)):

$$T_{2a} = T_1 + (T_2 - T_1)/\eta_{ic} \tag{3.3}$$

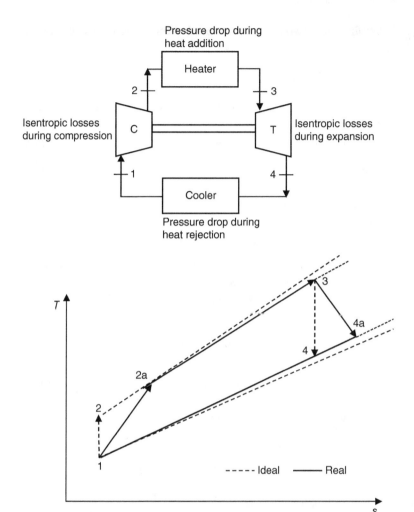

Figure 3.2 Real gas turbine processes.

Therefore, the actual work of compression in Equation 3.1 will become:

$$\dot{W}_c = -\dot{m}_a(h_{2a} - h_1) = \dot{m}_a C_{p,a}(T_{2a} - T_1)$$

3.3.3 The Combustion Chamber

The compressed gas enters a combustion chamber devoted to increasing the internal energy of the gas. Energy is supplied to the gas in the form of fuel, which is burnt at a given rate. The combustion process 2–3 is shown in Figure 3.2. The heat provided by combustion of the fuel is:

$$\dot{Q}_{in} = \dot{m}_{fuel} \times CV_{fuel} \times \eta_{comb} \tag{3.4}$$

This heat transferred to the gas increases its temperature from T_{2a} to T_3, therefore:

$$\dot{Q}_{in} = \dot{m}_g \times C_{p,g} \times (T_3 - T_{2a}) \tag{3.5}$$

It is possible to combine these two equations to predict the combustor's outlet temperature, T_3.

3.3.4 The Turbine Unit

A similar analogy to that used for the compressor is applied to the turbine, with its temperature dropping to T_{4a} at the outlet.

$$\dot{W}_t = \dot{m}_g \times C_{p,g} \times (T_3 - T_{4a}) \tag{3.6}$$

where $\dot{m}_g = \dot{m}_a + \dot{m}_{fuel}$ (3.7)

The isentropic turbine exit temperature is given by:

$$T_4 = T_3 \left(\frac{P_4}{P_3} \right)^{(n-1)/n} \tag{3.8}$$

Note that the actual exhaust temperature, T_{4a}, from the turbine will be above the frictionless, adiabatic exhaust temperature, T_4, as a result of irreversibilities in the turbine. The actual temperature is given by (from Equation (1.10)):

$$T_{4a} = T_3 - \eta_{it}(T_3 - T_4) \tag{3.9}$$

The net work output is given by the expansion work minus the work provided to the compressor:

$$\text{Net work output} = \dot{m}_g C_{p,g}(T_3 - T_{4a}) - \dot{m}_a C_{p,a}(T_{2a} - T_1) \tag{3.10}$$

3.3.5 Overall Performance of Gas Turbine Plants

- The compressor work, $\dot{W}_c$

$$\dot{W}_c = -\dot{m}_a C_{p,a}(T_{2a} - T_1) \tag{3.11}$$

- The heat provided by combustion of the fuel:

$$\dot{Q}_{in} = \dot{m}_{fuel} CV_f \eta_{comb} \tag{3.12}$$

or

$$\dot{Q}_{sup} = \dot{m}_a C_{p,a}(T_3 - T_{2a}) \tag{3.13}$$

- The turbine work, W_t

$$\dot{W}_t = \dot{m}_g C_{p,g}(T_3 - T_{4a}) \tag{3.14}$$

- Net work output:

$$= \dot{m}_g C_{p,g}(T_3 - T_{4a}) - \dot{m}_a C_{p,a}\left(T_{2a} - T_1\right) \tag{3.15}$$

- Thermal efficiency:

$$\eta_{cycle} = \frac{\text{Net work output}}{\text{Heat input}}$$

$$= \frac{\dot{m}_g C_{p,g}(T_3 - T_{4a}) - \dot{m}_a C_{p,a}(T_{2a} - T_1)}{\dot{m}_a C_{p,a}(T_3 - T_{2a})} \tag{3.16}$$

- Work ratio:

$$\text{Work ratio} = \frac{\text{Net work output}}{\text{Turbine work}}$$

$$WR = \frac{\dot{m}_g C_{p.g}(T_3 - T_{4a}) - \dot{m}_a C_{p.a}(T_{2a} - T_1)}{\dot{m}_g C_{p.g}(T_3 - T_{4a})} \tag{3.17}$$

3.4 Modifications to the Simple Gas Turbine Cycle

The performance of a gas turbine operating on the basic *Joule* cycle can be improved by employing one or more of the following modifications:

(a) Incorporating a heat exchanger.
(b) Splitting the compression process with 'intercooling' between stages.
(c) Splitting the expansion process with 'reheating' between stages.
(d) A compound system which incorporate all three modifications above.

These are discussed further in the following sections.

3.4.1 Heat Exchanger

The temperature of the exhaust gases leaving the turbine (T_{4a}) is higher than that at the end of the compression (T_{2a}) (see Figure 3.3). Hence, it is logical to reclaim the energy possessed by these gases before they are allowed to pass to the atmosphere. Thus, the energy consumption in the combustor is reduced and the thermal efficiency of the cycle is increased. A heat exchanger is a simple unit allowing the exchange of heat between the gases at T_{4a} with those at T_{2a}, resulting in a temperature rise of the compressed air to T_{2b} (see Figure 3.3). The heat exchanger effectiveness (η_{XE}) is defined as:

$$\eta_{XE} = \frac{\dot{m}_a C_{p.a}(T_{2b} - T_{2a})}{\dot{m}_g C_{p.g}(T_{4a} - T_{2a})} \tag{3.18}$$

If the mass flow of air is much higher than that of the fuel, as is usually the case in gas turbines (i.e. $\dot{m}_a \approx \dot{m}_g$) and assuming that differences in the specific heat capacity are very small (i.e. $C_{p.a} = C_{p.g}$) then the heat exchanger effectiveness (η_{XE}) may be given by the following simple expression:

$$\eta_{XE} = \frac{T_{2b} - T_{2a}}{T_{4a} - T_{2a}} \tag{3.19}$$

3.4.2 Intercooling

The compression process may be divided into two or more stages. If the gas between stages is cooled down to the original inlet temperature, the work input for a given pressure ratio is reduced, as indicated in Figure 3.4. Notice isentropic processes have been assumed for clarity. (Remember that isobars are diverging as *s* is increased.)

Hence, $\dot{W}_c$ first stage + $\dot{W}_c$ second stage < $\dot{W}_c$ single stage; if both compressors have the same isentropic efficiency then $T_4 = T_2$.

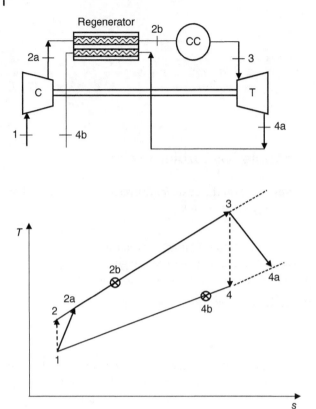

Figure 3.3 Regenerative gas turbine.

The intermediate pressure, P_I, for optimal work is given by differentiating the net work output (with respect to the intermediate pressure, P_I,) and equating to zero, resulting in:

$$P_I = \sqrt{P_L P_H} \tag{3.20}$$

where P_L and P_H are the low and high pressures, respectively.

3.4.3 Reheating

The expansion process, like compression, may also be performed in two or more turbine stages (see Figure 3.5). Usually, the high-pressure turbine (HPT) drives the compressor and the low-pressure turbine (LPT) provides the useful power output. This useful output is increased by reheating (RH) the gas before entering the LPT. By analogy with intercooling in the compressor, it can easily be shown that:

$\dot{W}_t$ first stage + $\dot{W}_t$ second stage > $\dot{W}_t$ single stage and again:

$$P_I = \sqrt{P_L P_H} \tag{3.21}$$

The intermediate pressure, P_I, produces the optimal power output, and $T_{4b} = T_{4a}$ if both stages have the same isentropic efficiencies.

Figure 3.4 Gas turbine with intercooling.

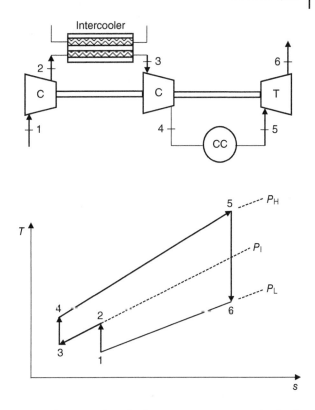

Using a reheat option increases the work output but the heat input is increased, which reduces the cycle efficiency. On the other hand, the temperature of the gases leaving the low-pressure turbine is relatively higher when reheating is used, and consequently a heat exchanger will produce better heat reclaim of the gases before they disappear to the atmosphere.

3.4.4 Compound System

Such a system includes all of the three modifications described above, as shown in Figure 3.6.

In order to find the optimal intermediate pressure, consider the gas turbine cycle shown in Figure 3.7 (isentropic processes assumed).

Since process 6–7 is reheating at constant pressure until $T_7 = T_5$, the turbine work can be found as:

$$\dot{W}_t = \dot{m}_g C_{p.g}(T_5 - T_6) + \dot{m}_g C_{p.g}(T_7 - T_8)$$
$$= \dot{m}_g C_{p.g} T_5 \left[1 + \frac{T_6}{T_5} + \frac{T_7}{T_5} - \frac{T_8}{T_5} \right]$$

But

$$\frac{T_6}{T_5} = \left(\frac{P_I}{P_H} \right)^{(n-1)/n} , \quad \frac{T_8}{T_7} = \left(\frac{P_L}{P_I} \right)^{(n-1)/n} , \quad T_5 = T_7$$

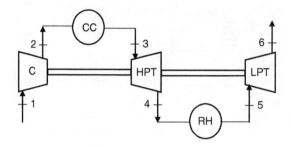

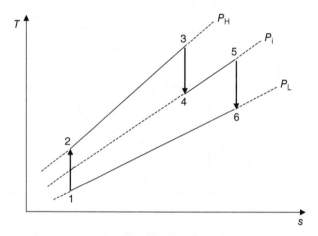

Figure 3.5 Gas turbine with reheating.

Therefore

$$\dot{W}_t = \dot{m}_g C_{p,\,g} T_5 \left[1 - \left(\frac{P_I}{P_H} \right)^{(n-1)/n} + 1 - \left(\frac{P_L}{P_I} \right)^{(n-1)/n} \right] \tag{3.22}$$

$$= \dot{m}_g C_{p,\,g} T_5 \left[2 - \left(\frac{P_I}{P_H} \right)^r - \left(\frac{P_L}{P_I} \right)^r \right]$$

where $r = (n-1)/n$

To find the maximum parameters:

$$\frac{d\dot{W}_t}{dp_I} = 0$$

$$= \dot{m}_g C_p T_5 [0 - r P_I^{r-1} P_H^{-r} + r P_L^r P_I^{-r-1}] = 0$$

Therefore

$$P_I^{r-1} I_i^{-(-r-1)} = P_L^r P_H^r$$

Hence

$$P_I - \sqrt{P_L P_H}$$

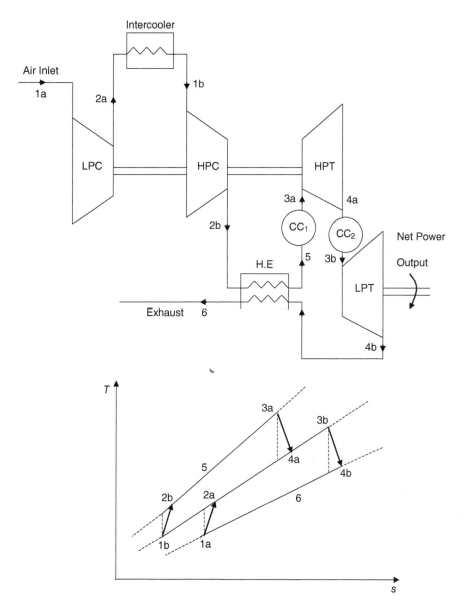

Figure 3.6 Compound gas turbine cycle.

3.4.5 Combined Gas Turbine/Steam Turbine Cycle

A combined-cycle power system typically uses a gas turbine to drive an electrical generator and recovers waste heat from the gas turbine exhaust to generate steam, which,
in turn, is used to drive a steam turbine, also connected to an electrical generator. The
overall electrical efficiency of a combined-cycle power system is, not surprisingly, higher
than the simple gas turbine cycle due to the utilization of otherwise waste heat. This type
of arrangement is sensible since gas turbines operate at higher temperatures than steam

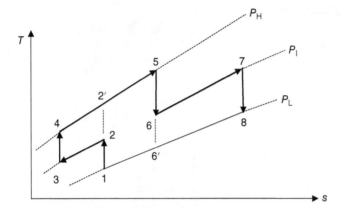

Figure 3.7 Optimal intermediate pressures for expansion and compression.

turbines due to the direct combustion process, and therefore the exit temperature of a gas turbine unit is usually about 500 °C, which has a fair amount of potential thermal energy to run a steam turbine Rankine cycle with moderate superheat condition.

In Figure 3.8, the Brayton gas turbine cycle processes are: 1–2–3–4, and the Rankine steam cycle processes are: 11–12–13–14.

Gas turbine calculations follow the normal Brayton cycle. Hence, it can be assumed that calculating the gas turbine cycle power output and cycle efficiency is an easy task. The waste heat recovery unit function is to allow the transfer of heat from the gas turbine unit to the steam turbine unit, converting water into steam from state 12 to state 13. Hence,

$$\dot{Q}_s = \dot{m}_g C_{pg}(T_4 - T_1) \times H_{lf} = \dot{m}_s(h_{13} - h_{12}) \tag{3.23}$$

where H_{lf} is the heat loss factor of the heat recovery unit, with a typical value of 0.95.

The process of estimation of the enthalpy values at points 11, 12, 13 and 14 was discussed in the previous relevant sections.

Once again, the calculations of the power output from the steam turbine and the steam cycle efficiency are straightforward.

The combined cycle calculation is the sum of the two cycles; hence, the combined cycle power output is:

$$\dot{W}_{comb} = \dot{W}_{netGT} + \dot{W}_{netST} \tag{3.24}$$

and the combined cycle efficiency is:

$$\eta_{comb} = \frac{\dot{W}_{GT} + \dot{W}_{ST}}{\dot{Q}_{2-3}}$$

$$= \frac{[C_{pg}(T_3 - T_4) - C_{pa}(T_2 - T_1)] + [(h_{13} - h_{14}) - (h_{12} - h_{11})]}{C_{pg}(T_3 - T_2)} \tag{3.25}$$

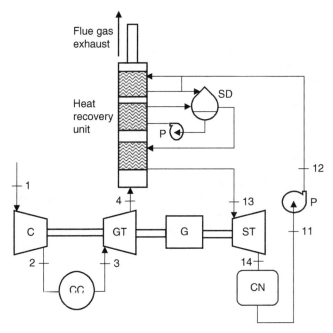

Key

C compressor	CC combustion chamber	GT gas turbine
G generator	ST steam turbine	CN condenser
P pump	SD steam drum	

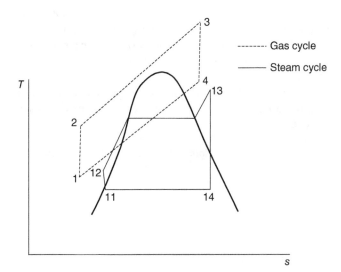

Figure 3.8 Combined gas turbine/steam turbine system.

3.5 Gas Engines

There are two categories of engine in use in the power industry. They are classified according to the supply of heat to produce the mechanical power that, in turn, is converted into electrical energy. These are:

Internal combustion engines: These use either the standard Otto cycle or the Diesel cycle.
External combustion engines: This type mainly relies on the Stirling engine.

3.5.1 Internal Combustion Engines

An internal combustion engine converts fuel into mechanical energy. The most common fuel is liquid, such as petrol (gasoline) or diesel. There are distinct differences in the design and operation of these two fuels and the engine design necessary to utilize them. There are also two thermodynamic cycles to describe the behaviour of these two engines.

The applications of internal combustion engines include the provision of the mechanical power source for most transport whether by road, rail, sea or air, as well as for industrial equipment such as air compressors and pumps, and for the provision of electrical energy, as in the case of portable power generators.

3.5.2 The Otto Cycle

This practical cycle applies to the petrol or spark-ignition engine, which is perhaps the most common heat engine in popular use. The cycle was originally proposed in 1862 but was made practicable by the German scientist Nikolaus Otto in 1876.

The Otto cycle is an ideal air standard cycle, which approximates the actual cycle. Figure 3.9 illustrates the Otto cycle on a *P–v* diagram.

The cycle consists of four non-flow processes.

In state 1, the engine cylinder is assumed to be full of air at approximately atmospheric pressure and temperature. The piston is at the bottom dead centre (BDC) position.

Process 1–2 is an isentropic (adiabatic and reversible) compression of the air. The piston moves to top dead centre (TDC), compressing the air into the clearance volume and so raising its pressure and temperature.

Process 2–3 is heat addition at constant volume. The piston remains at TDC whilst the heat is supplied from the surroundings and the pressure and temperature are raised to their maximum values in the cycle.

Figure 3.9 The Otto cycle *P–v* diagram.

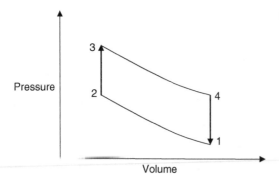

Pressure

Volume

Process 3–4 is an isentropic expansion. The hot high-pressure air forces the piston down the cylinder to BDC, thus producing useful work.

Process 4–1 is heat rejection at constant volume. The piston remains at BDC whilst the heat is transferred to the surroundings and the air returns to its original state at 1.

3.5.2.1 Analysis of the Otto Cycle

The air standard efficiency of the Otto cycle may be defined as:

$$\eta_{Otto} = \frac{\text{Net work output}}{\text{Heat supplied}} = \frac{w_{net}}{q_{in}}$$

If m is the mass of air in the cylinder and C_v is the specific heat capacity of air at constant volume, then:

Heat supplied: $q_{in} = C_v(T_3 - T_2)$

Heat rejected: $q_{out} = C_v(T_4 - T_1)$

Applying the first law of thermodynamics:

Net work output: $w_{net} = q_{in} - q_{out}$

It can be shown that the efficiency of the Otto cycle is:

$$\eta_{Otto} = \frac{w_{net}}{q_{in}} = \frac{q_{in} - q_{out}}{q_{in}} = 1 - \frac{q_{out}}{q_{in}} = 1 - \frac{(T_4 - T_1)}{(T_3 - T_2)} \tag{3.26}$$

It is often more convenient to express the air standard cycle efficiency in terms of the volume ratio. The compression ratio (r_v) is defined as:

$$r_v = V_1 / V_2$$

where V_1 = cylinder volume at BDC = swept volume + clearance volume.
V_2 = cylinder volume at TDC = clearance volume.
Using the process equations to replace temperatures by volumetric ratios:

$$\frac{T_2}{T_1} = \left(\frac{V_1}{V_2}\right)^{n-1} = r_v^{n-1}, \quad \frac{T_3}{T_4} = \left(\frac{V_4}{V_3}\right)^{n-1} = r_v^{n-1}$$

Using these definitions one can express the efficiency in terms of compression ratio only:

$$\eta_{Otto} = 1 - \frac{1}{r_v^{n-1}} \tag{3.27}$$

3.5.3 The Diesel Cycle

Rudolf Diesel invented the 'Diesel' cycle around 1892. This is the cycle used by the older type of diesel engine in which the fuel was injected into the cylinder by a blast of compressed air. The diesel fuel is injected into the cylinder at a temperature sufficient to initiate combustion; hence, the cycle is described as compression ignition and works at higher pressures than a petrol engine. Combustion therefore takes place at almost constant pressure, as opposed to the constant volume combustion of the Otto cycle. This difference in the heat supply process is the only difference in the air standard cycle for the two engines. Figure 3.10 shows the air standard Diesel cycle on a *P–v* diagram.

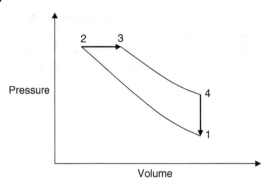

Figure 3.10 The Diesel cycle *P–v* diagram.

The cycle consists of four non-flow processes.

In state 1, the cylinder is assumed to be full of air at atmospheric pressure and temperature, and the piston is at BDC.

Process 1–2 is an isentropic compression of the air. The piston moves to TDC, compressing the air into the clearance volume and so raising its pressure and temperature.

Process 2–3 is heat addition at constant pressure. Heat is supplied to the air, resulting in a further increase in air temperature to its maximum at 3. This produces an increase in volume until the heat supply is cut off at 3.

Process 3–4 is an isentropic expansion of the air (product of combustion) during the remainder of the stroke until the piston reaches BDC at 4.

Process 4–1 is heat rejection at constant volume. The piston remains at BDC whilst the heat is transferred to the surroundings and the air returns to its original state at 1.

3.5.3.1 Analysis of the Diesel Cycle

The air standard Diesel cycle efficiency may be defined as:

$$\eta_{\text{Diesel}} = \frac{\text{Net work output}}{\text{Heat supplied}} = \frac{w_{\text{net}}}{q_{\text{in}}}$$

If m is the mass of air in the cylinder and the specific heat capacities of air at constant pressure and constant volume are C_p and C_v respectively, then:

Heat supplied $q_{\text{in}} = C_p(T_3 - T_2)$

Heat rejected $q_{\text{out}} = C_v(T_4 - T_1)$

Applying the first law of thermodynamics:

Net work output $w_{\text{net}} = q_{\text{in}} - q_{\text{out}}$

It can be shown that the efficiency of the Diesel cycle is:

$$\eta_{\text{Diesel}} = 1 - \frac{1}{n}\frac{(T_4 - T_1)}{(T_3 - T_2)} \tag{3.28}$$

An alternative expression for the air standard efficiency of the Diesel cycle may be determined from certain volume ratios for the cycle. As with the Otto cycle, the compression ratio is defined as:

$$r_v = V_1/V_2$$

The cut-off ratio is defined as:

$$r_c = V_3/V_2$$

Since the process 2–3 takes place at constant pressure:

$$r_c = \frac{V_3}{V_2} = \frac{T_3}{T_2}$$

Using the volume ratios, the air standard efficiency of the Diesel cycle becomes:

$$\eta_{\text{Diesel}} = 1 - \frac{1}{n}\frac{1}{r_v^{n-1}}\frac{(r_c^n - 1)}{(r_c - 1)} \tag{3.29}$$

Since n is a constant for air and r_v is a constant for a given engine construction, the cycle efficiency clearly depends on r_c. An increase in the load on the engine necessitates increasing r_c, which results in a decrease of efficiency, and this is one of the reasons why the Diesel cycle has been replaced for modern diesel engines by the dual combustion cycle.

3.5.4   The Dual Combustion Cycle

This cycle, sometimes known as the mixed cycle or semi-diesel cycle, closely approximates the working cycle of modern diesel engines. It is derived from an engine invented in the last quarter of the 19th century by British engineer Herbert Ackroyd-Stuart. It is, effectively, a combination of the Otto and Diesel cycles.

In the modern diesel engine, the fuel is forced into the cylinder in a high-pressure spray, with injection taking place both before and after the piston reaches top dead centre. Fuel injected before TDC ignites and burns instantaneously, producing a constant-volume combustion process. Subsequent fuel entering the cylinder burns as soon as it leaves the injection nozzle, with the piston moving down the cylinder again, thus producing a constant-pressure combustion process. The engine uses liquid fuel injection (i.e. unmixed liquid diesel oil) as opposed to the previous cycle in which the diesel was blown into the cylinder mixed with compressed air.

The cycle consists of five non-flow processes (see Figure 3.11).

In state 1, the cylinder is assumed to be full of air at approximately atmospheric pressure and temperature, and the piston is at the bottom dead centre (BDC) position.

Process 1–2 is an isentropic compression of the air. The piston moves to TDC, compressing the air into the clearance volume and so raising its pressure and temperature.

Figure 3.11 The dual combustion cycle P–v diagram.

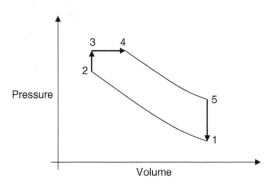

Process 2–3 is heat addition at constant volume supplied to the air, resulting in a further increase in both the pressure and temperature of the air.

Process 3–4 is heat addition at constant pressure. This causes an increase in volume and a further increase in the temperature of the air. The heat supply is cut off at 4 and the temperature of the air is at the maximum value attained during the cycle.

Process 4–5 is an isentropic expansion through the remainder of the stroke until the piston reaches BDC at 5.

Process 5–1 is heat rejection at constant volume. The piston remains at BDC whilst the heat is transferred to the surroundings and the air returns to its original state at 1.

3.5.4.1 Analysis of the Dual Cycle

If m is the mass of air in the cylinder and the specific heat capacities of air at constant pressure and constant volume are C_p and C_v respectively, then:

Heat supplied $q_{in} = C_v(T_3 - T_2) + C_p(T_4 - T_3)$

Heat rejected $q_{out} = C_v(T_5 - T_1)$

Applying the first law of thermodynamics:

Net work output $w_{net} = q_{in} - q_{out}$

Hence, the efficiency of the dual cycle is found to be:

$$\eta_{Dual} = 1 - \frac{(T_5 - T_1)}{(T_3 - T_2) + n(T_4 - T_3)} \tag{3.30}$$

An alternative expression for the air standard cycle efficiency can be derived in terms of certain volume and pressure ratios. However, the equation is rather too cumbersome to use. It is preferable to determine the work output and the heat supplied to evaluate the cycle efficiency, or to determine the temperature at each cycle point.

3.5.5 Diesel Engine Power Plants

Diesels are internal combustion reciprocating engines. They were developed primarily for marine propulsion and stationary service. The size of these units, in power output, is relatively small for power generation, but physically very large per unit of output, as compared to other types of generation.

They still have a place in smaller stationary power plant applications, performing emergency standby and peak load lopping duties. Additionally, they have the advantages of quick starting, relatively high efficiency and lower (CO_2) emissions. At the time of writing, however, their particulate and nitrogen oxide emission levels are coming under scrutiny.

The components of a typical diesel power plant are shown in Figure 3.12.

3.5.6 External Combustion Engines – The Stirling Engine

In a Stirling engine, a fixed amount of gas is enclosed in a working volume. The operation of a Stirling cycle is shown in Figure 3.13 on a P–v basis. It comprises two isothermal (1–2 and 3–4) and two isochoric (2–3 and 4–1) processes, which combine to form a closed working cycle. Heat addition takes place between 2 and 3.

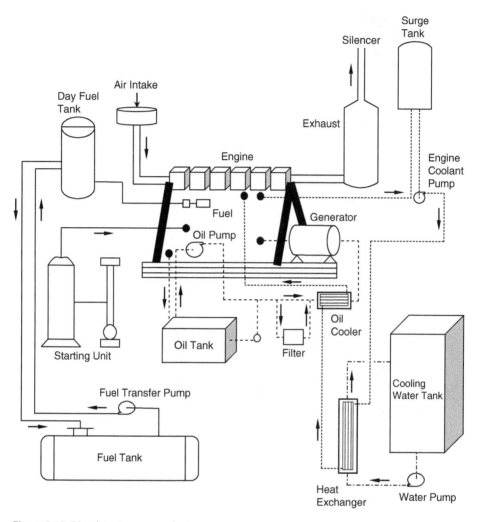

Figure 3.12 Diesel engine power plant.

The theoretical cycle efficiency of a Stirling engine is equal to that of the Carnot, and, like the Carnot, is impossible to achieve because ideal processes are impractical.

Applying the first law of thermodynamics:

$$\sum w = \sum q$$

$$q - w = \Delta u$$

For the isothermal compression process 1–2:

$$\Delta u = C_v \Delta T = 0$$

Hence

$$q_{12} = w_{12} = \int P dV = RT_c \ln\left(\frac{V_2}{V_1}\right)$$

3
T_h
4
Pressure
2
T_c
1
Volume

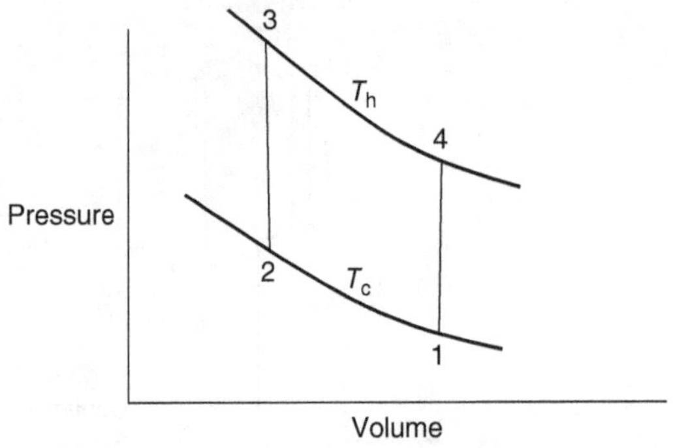

Figure 3.13 Ideal Stirling cycle.

Similarly, the process 3–4 is an isothermal expansion for which:

$$q_{34} = w_{34} = \int P\mathrm{d}V = RT_{\mathrm{h}} \ln\left(\frac{V_4}{V_3}\right)$$

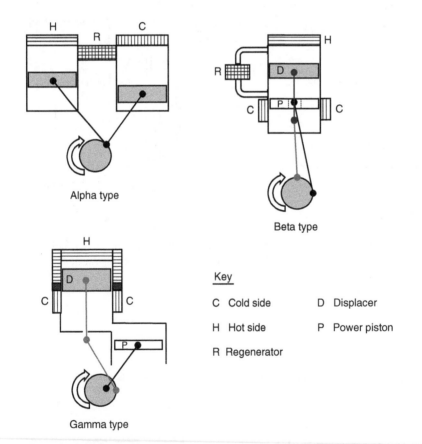

H
R
C
Alpha type
H
R
D
P
C
C
Beta type
H
D
C
C
P
Gamma type
Key
C Cold side
D Displacer
H Hot side
P Power piston
R Regenerator

Figure 3.14 Stirling engine types.

Hence, the efficiency can be written as:

$$\eta = \frac{w_{net}}{q_{in}} = \frac{q_{in} - q_{out}}{q_{in}} = 1 - \frac{T_{cold}}{T_{hot}} \qquad (3.31)$$

which is identical to the Carnot efficiency.

In a real Stirling cycle, a regenerator is used to remove heat during process 4–1 and return it during process 2–3. So, the net heat input, $q_{sup} = q_{23} - q_{41}$.

Since the heat input is lower, the cycle efficiency is improved by regeneration.

There are three types of Stirling engine configuration:

The *alpha type* has two cylinders containing expansion and compression spaces with a regenerator at the interface.

The *beta type* has one cylinder and a displacer piston that moves the working fluid back and forth between the hot and cold heat exchangers.

The *gamma type* has a displacer and a power piston, each working in their own separate cylinders (see Figure 3.14).

3.6 Worked Examples

Worked Example 3.1 – Simple gas turbine cycle with 100% isentropic efficiencies
A gas turbine unit has a pressure ratio of 9:1 and minimum and maximum cycle temperatures of 17 and 927 °C. The isentropic efficiencies of both the compressor and turbine are 100%. The combustion process is 100% efficient and the pressure drop is negligible. Assume the air flow rate through the compressor to be 1.0 kg/s and the air-to-fuel ratio to be 100:1. Take $C_p = 1.0$ kJ/kg K and $n = 1.4$ to be constant throughout. Calculate:

(a) The work of compression, expansion, the net power output and the work ratio.
(b) The heat supplied and the thermal efficiency of the plant.

Solution
Given: $P_2:P_1 = 9$, $T_1 = 290$ K, $T_3 = 1200$ K, $\eta_{ic} = 1$, $\eta_{it} = 1$, $\eta_{comb} = 1$, $\dot{m}_a = 1$ kg/s, $A:F = 100$, $C_p = 1$ kJ/kg K, $n = 1.4$ (see Figure 3.1).
Find $\dot{W}_c$, $\dot{W}_t$, $\dot{W}_{net}$, WR, $\dot{Q}_{in}$, η_{th}.
For the compressor:

$$T_2 = T_1 \left[\frac{P_2}{P_1}\right]^{\frac{n-1}{n}} = 290(9)^{\frac{0.4}{1.4}} = 543.3 \text{ K}$$

$$T_{2a} = T_1 + \frac{T_2 - T_1}{\eta_c} = 290 + \frac{543.3 - 290}{1} = 543.3 \text{ K}$$

For the turbine:

$$T_4 = T_3 \left[\frac{P_4}{P_3}\right]^{\frac{n-1}{n}} = 1200(1/9)^{\frac{0.4}{1.4}} = 640.5 \text{ K}$$

$$T_{4a} = T_3 - \eta_t(T_3 - T_4) = 1200 - 1(1200 - 640.5) = 640.5 \text{ K}$$

Compressor work input:

$$\dot{W}_c = \dot{m}_a C_{p,a}(T_{2a} - T_1) = 1.0 \times 1.0 \times (543.3 - 290) = 253 \text{ kW}$$

Turbine work output:

$$\dot{W}_t = \dot{m}_g C_{p,g}(T_3 - T_{4a}) = \left(1.0 + \frac{1}{100}\right) \times 1.0 \times (1200 - 640.5) = 565 \text{ kW}$$

Net work output $\dot{W}_{net} = \dot{W}_t - \dot{W}_c = 565 - 253 = 312$ kW

$\dot{Q}_{in} = \dot{m}_a C_{p,a}(T_3 - T_{2a}) = 1.0 \times 1.0 \times (1200 - 640.5) = 657$ kW

Cycle efficiency $= \dfrac{\dot{W}_{net}}{\dot{Q}_{in}} = \dfrac{312}{657} \times 100 = 47\%$

Work ratio $= \dfrac{\dot{W}_{net}}{\dot{W}_t} = \dfrac{312}{565} = 0.552$

Worked Example 3.2 – Simple gas turbine cycle with less than 100% isentropic efficiencies

A gas turbine unit has a pressure ratio of 9:1 and minimum and maximum cycle temperatures of 17 and 927 °C. The isentropic efficiencies of the compressor and turbine are 80% and 90%, respectively. The combustion process is 100% efficient and the pressure drop is negligible. Assume the air flow rate through the compressor to be 1.0 kg/s and the air-to-fuel ratio to be 100:1. Take $C_p = 1.0$ kJ/kg K and $n = 1.4$ to be constant throughout. Calculate:

(a) The work of compression, expansion, the net power output and the work ratio.
(b) The heat supplied and the thermal efficiency of the plant.

Solution

Given: $P_2 : P_1 = 9$, $T_1 = 290$ K, $T_3 = 1200$ K, $\eta_{ic} = 0.8$, $\eta_{it} = 0.9$, $\eta_{comb} = 1$, $\dot{m}_a = 1$ kg/s, $A : F = 100$, $C_p = 1$ kJ/kg K, $n = 1.4$ (see Figure 3.2).
Find $\dot{W}_c$, $\dot{W}_t$, $\dot{W}_{net}$, WR, $\dot{Q}_{in}$, η_{th}.

$T_2 = T_1 \left[\dfrac{P_2}{P_1} \right]^{\frac{n-1}{n}} = 290(9)^{\frac{0.4}{1.4}} = 543.3$ K

$T_{2a} = T_1 + \dfrac{T_2 - T_1}{\eta_c} = 290 + \dfrac{543.3 - 290}{0.8} = 606.6$ K

$T_4 = T_3 \left[\dfrac{P_4}{P_3} \right]^{\frac{n-1}{n}} = 1200(1/9)^{\frac{0.4}{1.4}} = 640.5$ K

$T_{4a} = T_3 - \eta_t(T_3 - T_4) = 1200 - 0.9(1200 - 640.5) = 696.5$ K

$\dot{W}_c = \dot{m}_a C_p(T_{2a} - T_1) = 1.0 \times 1.0 \times (606.6 - 290) = 316.6$ kW

$\dot{W}_t = \dot{m}_g C_{p.g}(T_3 - T_{4a}) = \left(1.0 + \dfrac{1}{100}\right) \times 1.0 \times (1200 - 696.5) = 508.6$ kW

Net work output $\dot{W}_{net} = \dot{W}_t - \dot{W}_c = 508.6 - 316.6 = 191.9$ kW

$\dot{Q}_{in} = \dot{m}_a C_{p,a}(T_3 - T_{2a}) = 1.0 \times 1.0 \times (1200 - 607.6) = 593.4$ kW

Cycle efficiency $= \dfrac{\dot{W}_{net}}{\dot{Q}_{in}} = \dfrac{191.9}{593.4} \times 100 = 32.35\%$

Work ratio $= \dfrac{\dot{W}_{net}}{\dot{W}_t} = \dfrac{191.9}{508.6} = 0.377$

Worked Example 3.3 – Simple gas turbine cycle with less than 100% isentropic efficiencies and accounting for pressure drop in the combustor

A gas turbine unit has a pressure ratio of 9:1 and minimum and maximum cycle temperatures of 17 and 927 °C. The isentropic efficiencies of the compressor and turbine are 80% and 90%, respectively. The combustion process is 100% efficient and the pressure drop in the combustion chamber is 0.1 bar. Assume the air flow rate through the compressor to be 1.0 kg/s and the air-to-fuel ratio to be 100:1. Take $C_p = 1.0$ kJ/kg K and $n = 1.4$ to be constant throughout. Calculate:

(a) The work of compression, expansion, the net power output and the work ratio.
(b) The heat supplied and the thermal efficiency of the plant.

Solution

Given: $P_2 : P_1 = 9$, $T_1 = 290$ K, $T_3 = 1200$ K, $\eta_{ic} = 0.8$, $\eta_{it} = 0.9$, $\eta_{comb} = 1$, $\dot{m}_a = 1$ kg/s, $A : F = 100$, $C_p = 1$ kJ/kg K, $n = 1.4$ (see Figure 3.2).

Find $\dot{W}_c$, $\dot{W}_t$, $\dot{W}_{net}$, WR, $\dot{Q}_{in}$, η_{th}.

$$T_2 = T_1 \left[\frac{P_2}{P_1} \right]^{\frac{n-1}{n}} = 290(9)^{\frac{0.4}{1.4}} = 543.3 \text{ K}$$

$$T_{2a} = T_1 + \frac{T_2 - T_1}{\eta_c} = 290 + \frac{543.3 - 290}{0.8} = 606.6 \text{ K}$$

$$T_4 = T_3 \left[\frac{P_4}{P_3} \right]^{\frac{n-1}{n}} = 1200(1/8.9)^{\frac{0.4}{1.4}} = 642.6 \text{ K}$$

$$T_{4a} = T_3 - \eta_t(T_3 - T_4) = 1200 - 0.9(1200 - 640.5) = 698.3 \text{ K}$$

$$\dot{W}_c = \dot{m}_a C_{p,a}(T_{2a} - T_1) = 1.0 \times 1.0 \times (606.6 - 290) = 316.6 \text{ kW}$$

$$\dot{W}_t = \dot{m}_g C_{p,g}(T_3 - T_{4a}) = \left(1.0 + \frac{1}{100} \right) \times 1.0 \times (1200 - 698.3) = 507 \text{ kW}$$

Net work output $\dot{W}_{net} = \dot{W}_t - \dot{W}_c = 507 - 316.6 = 190 \text{ kW}$

$$\dot{Q}_{in} = \dot{m}_a C_{p,a}(T_3 - T_{2a}) = 1.0 \times 1.0 \times (1200 - 607.6) = 593.4 \text{ kW}$$

$$\text{Cycle efficiency} = \frac{\dot{W}_{net}}{\dot{Q}_{in}} = \frac{190}{593.4} \times 100 = 32.03\%$$

$$\text{Work ratio} = \frac{\dot{W}_{net}}{\dot{W}_t} = \frac{190}{507} = 0.375$$

Worked Example 3.4 – Gas turbine cycle with irreversibilities and heat exchanger

The gas turbine plant in Worked Example 3.1 is modified to include a heat exchanger to heat the compressed air by the gases leaving the turbine unit. If the heat exchanger has an effectiveness of 70%, estimate the increase in cycle efficiency achieved by its introduction. Take $n = 1.4$ and $C_p = 1$ kJ/kg K constant through the cycle.

Solution

Given: $P_2 : P_1 = 9$, $T_1 = 290$ K, $T_3 = 1200$ K, $\eta_{ic} = 0.8$, $\eta_{it} = 0.9$, $\eta_{comb} = 1$, $\dot{m}_a = 1$ kg/s, $A : F = 100$, $C_p = 1$ kJ/kg K, $n = 1.4$, $\eta_{XE} = 0.7$ (see Figure 3.3).

Find $\dot{W}_c$, $\dot{W}_t$, $\dot{W}_{net}$, WR, $\dot{Q}_{in}$, η_{th}.

The cycle original temperatures remain as in Worked Example 3.1, so the work of compression and the turbine output are unchanged, i.e.

$$T_2 = T_1 \left[\frac{P_2}{P_1} \right]^{\frac{n-1}{n}} = 290(9)^{\frac{0.4}{1.4}} = 543.3 \text{ K}$$

$$T_{2a} = T_1 + \frac{T_2 - T_1}{\eta_c} = 290 + \frac{543.3 - 290}{1} = 543.3 \text{ K}$$

$$T_4 = T_3 \left[\frac{P_4}{P_3} \right]^{\frac{n-1}{n}} = 1200(1/9)^{\frac{0.4}{1.4}} = 640.5 \text{ K}$$

$$T_{4a} = T_3 - \eta_t(T_3 - T_4) = 1200 - 1(1200 - 640.5) = 640.5 \text{ K}$$

Compressor work input:

$$\dot{W}_c = \dot{m}_a C_{p,a}(T_{2a} - T_1)1 = 1.0 \times 1.0 \times (543.3 - 290) = 253 \text{ kW}$$

Turbine work output:

$$\dot{W}_t = \dot{m}_g C_{p,g}(T_3 - T_{4a}) = \left(1.0 + \frac{1}{100} \right) \times 1.0 \times (1200 - 640.5) = 565 \text{ kW}$$

Net work output $\dot{W}_{net} = \dot{W}_t - \dot{W}_c = 565 - 253 = 312 \text{ kW}$

$$\text{Work ratio} = \frac{\dot{W}_{net}}{\dot{W}_t} = \frac{312}{565} = 0.552$$

The difference in the regenerative cycle is the temperature at inlet to the combustor. The heat supplied will change as a consequence:

$$T_{2b} = T_{2a} + \eta_{he}(T_{4a} - T_{2a}) = 543.3 + 0.7(640.5 - 543.3) = 611 \text{ K}$$

The heat supplied is reduced due to the increase in temperature of gases leaving the compressor:

$$\dot{Q}_{in} = \dot{m}_a C_{p,a}(T_3 - T_{2b}) = 1.0 \times 1.0 \times (1200 - 611) = 589 \text{ kW}$$

Therefore, the cycle efficiency will change:

$$\eta_{cycle} = \frac{\dot{W}_{net}}{\dot{Q}_{in}} = \frac{312}{589} \times 100 = 52.96\%$$

Hence, regeneration (heat reclaim) has improved the cycle efficiency by 6%.

Worked Example 3.5 – Two-stage expansion with irreversibilities

A gas turbine unit takes in air at 17 °C. The air flow rate is 1 kg/s through a pressure ratio of 9:1 and the maximum cycle temperature is 927 °C. The compression takes place in one stage, while expansion is split into two stages with optimal condition intermediate pressure and reheating at the intermediate pressure to 927 °C. The isentropic efficiency of the compressor, HPT and LPT turbines is 100%. Assume the combustion process to be 100% efficient and losses of pressure to be negligible. Assume the air-to-fuel ratio is 100:1. Take $C_p = 1.0$ kJ/kg K and $n = 1.4$ to be constant throughout. Calculate the net power output, the cycle efficiency and the work ratio of the plant. Compare the results with those for a single-stage turbine.

Solution

Given: $P_2/P_1 = 9$, $T_1 = 290$ K, $T_3 = 1200$ K, $\eta_{ic} = \eta_{it} = \eta_{comb} = 1$, $\dot{m}_a = 1$ kg/s, $A:F = 100$, $C_p = 1$ kJ/kg K, $n = 1.4$, $\eta_{XE} = 0.7$ (see Figure 3.5).

Find $\dot{W}_c$, $\dot{W}_t$, $\dot{W}_{net}$, WR, $\dot{Q}_{in}$, η_{th}.

$$T_2 = T_1 \left[\frac{P_2}{P_1}\right]^{\frac{n-1}{n}} = 290(9)^{\frac{0.4}{1.4}} = 543.3 \text{ K}$$

$$T_{2a} = T_1 + \frac{T_2 - T_1}{\eta_c} = 290 + \frac{543.3 - 290}{1.0} = 543.3 \text{ K}$$

For optimal inter-staging:

$$P_4 = \sqrt{P_3 \times P_5} = \sqrt{9/1} = 3 \text{ bar}$$

$$T_4 = T_3 \left[\frac{P_4}{P_3}\right]^{\frac{n-1}{n}} = 1200(3/9)^{\frac{0.4}{1.4}} = 876.7 \text{ K}$$

$$T_{4a} = T_3 - \eta_t(T_3 - T_4) = 1200 - 1.0(1200 - 876.7) = 876.7 \text{ K}$$

$$T_6 = T_5 \left[\frac{P_5}{P_6}\right]^{\frac{n-1}{n}} = 1200(1/3)^{\frac{0.4}{1.4}} = 876.7 \text{ K}$$

$$T_{6a} = T_5 - \eta_t(T_5 - T_6) = 1200 - 1.0(1200 - 876.7) = 876.7 \text{ K}$$

Compressor work:

$$\dot{W}_c = \dot{m}_a C_{p,a}(T_2 - T_1)1 = 1 \times 1.0 \times (543.3 - 290) = 253 \text{ kW}$$

Turbine work:

$$\dot{W}_t = \dot{m}_g C_{p,g}[(T_3 - T_4) + (T_5 - T_6)]$$
$$= (1 + 1/100) \times 1.0 \times [(1200 - 876.7) + (1200 - 876.7)] = 653 \text{ kW}$$

Net work output $\dot{W}_{net} = \dot{W}_t - \dot{W}_c = 653 - 253 = 400$ kW

Work ratio $= \dfrac{\dot{W}_{net}}{\dot{W}_t} = \dfrac{400}{653} = 0.612$

Heat supplied, $\dot{Q}_{in} = \dot{m}_a C_{p,a}(T_3 - T_2) + \dot{m}_a C_{p,a}(T_5 - T_4)$

$$= \left(\left(1 + \frac{1}{100}\right) \times 1.0 \times (1200 - 543.3)\right)$$
$$+ (1 \times 1.0 \times (1200 - 876.7))$$
$$= 980 \text{ kW}$$

Cycle efficiency $= \dfrac{\dot{W}_{net}}{\dot{Q}_{in}} = \dfrac{400}{980} \times 100 = 40.8\%$

The efficiency achieved by splitting the turbine into two stages has increased from 32 to 40.8%.

Worked Example 3.6 – Otto cycle

An engine working on the Otto cycle has a compression ratio of 9:1. At the beginning of the compression stroke the temperature of the air is 27 °C and the pressure is 1 bar. 500 kJ/kg of heat is added at constant volume per cycle. Determine:

(a) The temperature at the remaining points in the cycle.
(b) The efficiency of the cycle.

For air, assume $n = 1.4$ and $C_v = 0.718$ kJ/kg K.

Solution

Given: $r_v = 9$, $T_1 = 300$ K, $P_1 = 1$ bar, $q_{in} = 500$ kJ/kg, $n = 1.4$, $C_v = 0.718$ kJ/kg K (see Figure 3.9).

Find T_2, T_3, T_4, η_{Otto}.

(a) $T_2 = T_1 \times r_v^{n-1} = 300 \times 9^{1.4-1} = 722.5$ K

Heat supplied at constant volume:

$q_{in} = C_v(T_3 - T_2)$

$500 = 0.718(T_3 - 722.5)$

$T_3 = 1418$ K

$T_4 = T_3/r_v^{n-1} = 1418/9^{0.4} = 589$ K

(b) Heat rejected: $q_{out} = C_v(T_4 - T_1) = 0.718(589 - 300) = 205.4$ kJ/kg

Net work done $w_{net} = q_{in} - q_{out} = 500 - 205.4 = 294.6$ kJ/kg

and

η_{Otto} = Net work done/Heat supplied = 294.6/500 = 0.589 or 58.9%

Alternatively:

$$\eta_{Otto} = 1 - \frac{1}{r_v^{n-1}} = 1 - \frac{1}{9^{1.4-1}} \times 100 = 58.47\%$$

Worked Example 3.7 – Diesel cycle

An air standard Diesel cycle has a compression ratio of 12:1. 500 kJ/kg of heat is supplied to the working fluid per cycle. At the beginning of the compression process, the pressure and temperature are 1.0 bar and 27 °C, respectively. Determine:

(a) The temperature at the remaining points in the cycle.
(b) The air standard thermal efficiency.

For air, assume $n = 1.4$; $C_p = 1.005$ kJ/kg K: $C_v = 0.718$ kJ/kg K.

Solution

Given: $r_v = 12$, $T_1 = 300$ K, $P_1 = 1$ bar, $q_{in} = 500$ kJ/kg, $n = 1.4$, $C_p = 1.005$ kJ/kg K, $C_v = 0.718$ kJ/kg K (see Figure 3.10).

Find T_2, T_3, T_4, η_{Diesel}.

(a) $T_2 = T_1 \times r_v^{n-1} = 300 \times 12^{1.4-1} = 810$ K

Process 2–3 is constant-pressure heating:

$q_{in} = C_p(T_3 - T_2)$

$500 = 1.005(T_3 - 810)$

Hence

$$T_3 = 1308 \text{ K}$$

Since $P_2 = P_3$:

$$V_3/V_2 = T_3/T_2 = 1308/810 = 1.6148$$

Process 3–4 is adiabatic expansion, hence:

$$T_4 = T_3(V_3/V_4)^{n-1}$$

But, since $V_4 = V_1$

$$V_3/V_4 = (V_3/V_2) \times (V_2/V_1) = 1.6148 \times 1/12 = 0.134$$

Hence

$$T_4 = 1308(0.134)^{0.4} = 586 \text{ K}$$

(b) Heat rejected:

$$q_{out} = C_v(T_4 - T_1) = 0.718(586 - 300) = 205.5 \text{ kJ/kg}$$

Net work done, $w_{net} = q_{in} - q_{out} = 500 - 205.5 = 294.5 \text{ kJ/kg}$

and

$$\eta_{Diesel} = \text{Net work done/Heat supplied} = 294.5/500 = 0.589 \text{ or } 58.9\%$$

Alternatively:

$$\eta_{Diesel} = 1 - \frac{1}{n} \times \frac{r_c^n - 1}{r_c - 1} \times \frac{1}{r_v^{n-1}} = 1 - \frac{1}{1.4} \times \frac{1.6148^{1.4} - 1}{1.6148 - 1} \times \frac{1}{12^{1.4-1}} \times 100 = 58.9\%$$

Worked Example 3.8 – Dual cycle diesel engine

A diesel engine works on the dual combustion cycle and has a compression ratio of 12:1. At the start of compression, the air is at a temperature of 27 °C and the pressure is 1 bar. In the cycle, 250 kJ/kg heat is added at constant volume and a further 250 kJ/kg at constant pressure. Calculate:

(a) The temperature at the remaining points in the cycle.
(b) The air standard efficiency of the cycle.

For air, assume $n = 1.4$; $C_p = 1.005 \text{ kJ/kg K}$; $C_v = 0.718 \text{ kJ/kg K}$.

Solution

Given: $r_v = 12$, $T_1 = 300 \text{ K}$, $P_1 = 1 \text{ bar}$, $q_{in,total} = 250 + 250 = 500 \text{ kJ/kg}$, $n = 1.4$, $C_p = 1.005 \text{ kJ/kg K}$, $C_v = 0.718 \text{ kJ/kg K}$ (see Figure 3.11).
Find $T_{2a}, T_2, T_3, T_4, \eta_{Dual\ Diesel}$.

(a) $T_{2a} = T_1 \times r_v^{n-1}$

$$= 300 \times 12^{1.4-1} = 810.58 \text{ K}$$

Process 2a–2 is constant-volume heating:

$$q_{in} = C_v(T_2 - T_{2a})$$

$$250 = 0.718(T_2 - 810.58)$$

Hence

$$T_2 = 1158.77 \text{ K}$$

Process 2–3 is constant-pressure heating:

$$q_{in} = C_p(T_3 - T_2)$$
$$250 = 1.005(T_3 - 1158.77)$$

Hence

$$T_3 = 1407.5 \text{ K}$$

Since $P_2 = P_3$:

$$V_3/V_2 = T_3/T_2 = 1407.5/1158.77 = 1.215$$

Process 3–4 is adiabatic expansion, hence:

$$T_4 = T_3(V_3/V_4)^{n-1}$$

But, since $V_4 = V_1$

$$V_3/V_4 = (V_3/V_2) \times (V_2/V_1) = 1.215 \times 1/12 = 0.101$$

Hence

$$T_4 = 1158.77(0.101)^{0.4} = 563 \text{ K}$$

(b) Heat rejected: $q_{out} = C_v(T_4 - T_1) = 0.718(563 - 300) = 188.9 \text{ kJ/kg}$

Net work done, $w_{net} = q_{in} - q_{out} = 500 - 188.99 = 311 \text{ kJ/kg}$

and

$$\eta_{Dual\ Diesel} = \text{Net work done/Heat supplied} = 311/500 = 0.622 \text{ or } 62.2\%$$

Worked Example 3.9 – Stirling engine

A Stirling engine uses helium as the working fluid, operating with an isothermal compression at a temperature $T_C = 20\ °C$ and an isothermal expansion process at temperature $T_h = 200\ °C$. It has a maximum volume of 650 cm³ and a minimum volume of 550 cm³. The lowest pressure in the cycle is 1 bar.

Estimate the cycle efficiency (%) and the net power output (W) when the engine speed is 50 cycles/s.

For helium, take $R = 2.077 \text{ kJ/kg K}$ and $C_v = 3.116 \text{ kJ/kg K}$.

Solution

Given: $T_c = 20\ °C = 293 \text{ K}$, $T_h = 200\ °C = 473 \text{ K}$, $V_2 = 650 \times 10^{-6} \text{ m}^3$, $V_1 = 550 \times 10^{-6} \text{ m}^3$, $C_v = 3.116 \text{ kJ/kg K}$, $R = 2.077 \text{ kJ/kg K}$, $N = 50 \text{ cycles/s}$ (see Figure 3.13).
Find η_{cycle}, $\dot{W}_{net}$

The cycle efficiency is given by:

$$\eta_{cycle} = 1 - \frac{T_c}{T_h} = 1 - \frac{20 + 273}{200 + 273} = 0.38 \text{ or } 38\%$$

In order to calculate the net power output, the mass has to be calculated:

$$m = \frac{PV}{RT} = \frac{1 \times 10^5 \times 650 \times 10^{-6}}{2077(20 + 273)} = 0.000107 \text{ kg/cycle}$$

$$W_{net} = W_{34} - W_{12} = mRT_h \ln\left(\frac{V_4}{V_3}\right) - mRT_c \ln\left(\frac{V_2}{V_1}\right)$$

$$= mR(T_h - T_c).\ln\frac{V_1}{V_2} = 0.000107 \times 2077 \times (200 - 20) \ln\left(\frac{650}{550}\right)$$

$$= 6.68 \text{ J/cycle}$$

Hence, the power output in kW = J/cycle x speed in cycle/s

$$\dot{W}_{net} = 6.68 \times 50 = 334 \text{ W}$$

Worked Example 3.10 – A combined gas turbine/steam turbine cycle
A gas turbine unit has a pressure ratio of 9:1 and minimum and maximum cycle temperatures of 17 and 927 °C. The isentropic efficiencies of the compressor and turbine are both 100%. The combustion process is 100% efficient and the pressure drop is negligible. Assume the air flow rate through the compressor to be 10 kg/s and the air-to-fuel ratio to be 100:1. Take $C_p = 1.0$ kJ/kg K and $n = 1.4$ to be constant throughout.

A heat recovery/steam generation unit is added to the gas turbine unit above to generate steam to feed an ideal Rankine superheat steam cycle operating between 10 kPa and 6 MPa with steam flow rate of 1 kg/s. If the steam generation/heat recovery unit is 80% efficient, calculate:

(a) The steam cycle net work output.
(b) The combined cycle efficiency.

Solution
Given: $P_2 : P_1 = 9$, $T_1 = 290$ K, $T_3 = 1200$ K, $\eta_{ic} = 1$, $\eta_{it} = 1$, $\eta_{comb} = 1$, $\dot{m}_a = 10$ kg/s, $\dot{m}_s = 1$ kg/s, $P_{11} = 10 \times 10^3$ Pa, $P_{12} = 6 \times 10^6$ Pa, $A : F = 100$, $C_p = 1$ kJ/kg K, $n = 1.4$, $\eta_{XE} = 0.8$ (see Figure 3.8).
Find $\dot{W}_{net,s}$, $\eta_{combined\ cycle}$.

$$T_2 = T_1 \left[\frac{P_2}{P_1}\right]^{\frac{n-1}{n}} = 290(9)^{\frac{0.4}{1.4}} = 543.3 \text{ K}$$

$$T_{2a} = T_1 + \frac{T_2 - T_1}{\eta_c} = 290 + \frac{543.3 - 290}{1} = 543.3 \text{ K}$$

$$T_4 = T_3 \left[\frac{P_4}{P_3}\right]^{\frac{n-1}{n}} = 1200(1/9)^{\frac{0.4}{1.4}} = 640.5 \text{ K}$$

$$T_{4a} = T_3 - \eta_t(T_3 - T_4) = 1200 - 1(1200 - 640.5) = 640.5 \text{ K}$$

Compressor work input:

$$\dot{W}_c = \dot{m}_a C_{p,a}(T_{2a} - T_1) = 10 \times 1.0 \times (543.3 - 290) = 2530 \text{ kW}$$

Turbine work output:

$$\dot{W}_t = \dot{m}_g C_{p.g}(T_3 - T_{4a}) = \left(10 + \frac{1}{100}\right) \times 1.0 \times (1200 - 640.5) = 5650 \text{ kW}$$

Net work output, $\dot{W}_{net} = \dot{W}_t - \dot{W}_c = 5650 - 2530 = 3120 \text{ kW}$

$$\dot{Q}_{in} = \dot{m}_a C_{p.a}(T_3 - T_{2a}) = 10 \times 1.0 \times (1200 - 640.5) = 6570 \text{ kW}$$

$$\text{Cycle efficiency} = \frac{\dot{W}_{net}}{\dot{Q}_{in}} = \frac{3120}{6570} \times 100 = 47\%$$

$$\dot{Q}_{rejected} = \dot{m}_a C_{p.a}(T_4 - T_1) = 10 \times 1.0 \times (640.5 - 290) = 3505 \text{ kW}$$

Assuming ideal conditions,

$$\dot{Q}_s = 0.8 \times (h_{13} - h_{12}) = 10 \times 10 \times (640.5 - 290) = 3505 \text{ kW}$$

But $\dot{Q}_s = 0.8 \times \dot{m}_s(h_{13} - h_{12})$, since the recovery unit is 80% efficient.

From this equation, it will be possible to determine the enthalpy of steam input to the steam turbine unit (h_{13}).

The steam cycle enthalpies are calculated as follows:

$h_{11} = 191.83 \text{ kJ/kg}$
$h_{12} = 191.83 + 0.001(6000 - 10) = 197.88 \text{ kJ/kg}$

Hence:

$$h_{13} = h_{12} + (\dot{Q}_s/\dot{m}_s) = 197.88 + \left(0.8 \times \frac{3505}{1}\right) = 3001.9 \text{ kJ/kg}$$

The state of steam at this point is superheated, at about 340 °C.
$s_{13} = 6.263 \text{ kJ/kg K}$
The expansion in the steam turbine from 6 MPa down to 10 kPa, at 100% isentropic leads to a final condition of steam:
$h_{14} = 1982.7 \text{ kJ/kg}$
Hence, the net work output for the steam cycle is:

$$\dot{W}_{net} = \dot{m}_s[(h_{13} - h_{14}) - (h_{12} - h_{11})]$$

$$= 1[(3001.9 - 1982.7) - (197.88 - 191.83)] = 1013 \text{ kW}$$

Therefore, the combined cycle efficiency is:

$$\eta_{combined cycle} = \frac{\dot{W}_{gas cycle} + \dot{W}_{steam cycle}}{\dot{Q}_{in}} = (3210 + 1013)/6570 = 0.64 \text{ or } 64\%$$

which is 17% higher than the GT cycle efficiency.

3.7 Tutorial Problems

3.1 A gas turbine unit has a pressure ratio of 6:1 and minimum and maximum cycle temperatures of 27 °C and 927 °C, respectively. The air enters the compressor at the rate of 1 kg/s and the combustor has an air-to-fuel ratio of 100:1. The isentropic efficiencies of the compressor and turbine are 100%.

Assume combustion efficiency is 100% and the pressure drop in the combustor is negligible. Take $C_p = 1.0$ kJ/kg K and $n = 1.4$ to be constant throughout. Calculate:

(a) The net work output.
(b) The cycle efficiency.
(c) The work ratio of the plant.

[Answers: 285 kW, 40.75%, 0.587]

3.2 Calculate the effect of isentropic efficiencies on the overall cycle efficiency of the plant in Problem 3.1. Take the isentropic efficiencies of the compressor and turbine to be 80% and 90%, respectively. Assume combustion efficiency is 100% and the pressure drop in the combustor is negligible.

[Answers: 186 kW, 28.70%, 0.426]

3.3 Calculate the effect of a pressure drop in the combustor on the cycle efficiency of the system described in Problem 3.1. Consider the pressure drop to be 10%.

[Answer: 263 kW, 37.58%, 0.567]

3.4 The gas turbine unit in Problem 3.1 is now fitted with a heat exchanger, with an effectiveness of 0.7. Calculate the reduction in the heat supply needed and the change in cycle efficiency.

[Answers: 153 kW, +12%]

3.5 A gas turbine unit takes in air at 27 °C, has an air flow of 1 kg/s through a pressure ratio of 16:1 and a maximum cycle temperature of 927 °C. The compression takes place in one stage, while expansion is split into two stages, with optimal condition intermediate pressure and reheating at the intermediate pressure to 927 °C. The isentropic efficiency of the compressor, HPT and LPT turbines is 100%. Assume the combustion process to be 100% efficient and losses in pressure to be negligible. Assume the air-to-fuel ratio is 100:1. Take $C_p = 1.0$ kJ/kg K and $n = 1.4$ to be constant throughout. Calculate the net power output and the cycle efficiency of the plant.

[Answers: 430 kW, 46%]

3.6 An engine working on the Otto cycle has a compression ratio of 9:1. At the beginning of the compression stroke, the temperature of the air is 0 °C and the pressure is 1 bar. 500 kJ/kg of heat is added at constant volume per cycle. Determine:

(a) The temperature at the remaining points in the cycle.
(b) The efficiency of the cycle.
For air, assume $n = 1.4$ and $C_v = 0.718$ kJ/kg K.

[Answers: (a) $T_2 = 657$ K, $T_3 = 1354$ K, $T_4 = 562$ K; (b) 58.47%]

3.7 An air standard Diesel cycle has a compression ratio of 9:1. 500 kJ/kg of heat is supplied to the working fluid per cycle. At the beginning of the compression process, the pressure and temperature are 1.0 bar and 0 °C, respectively. Determine the air standard thermal efficiency.

For air, assume $n = 1.4$, $C_p = 1.005$ kJ/kg K, $C_v = 0.718$ kJ/kg K.

[Answer: 52.9%]

3.8 A diesel engine works on the dual combustion cycle and has a compression ratio of 9:1. At the start of compression, the air is at a temperature of 0 °C and the pressure is 1 bar. In the cycle, 250 kJ/kg heat is added at constant volume and a further 250 kJ/kg at constant pressure. Calculate the air standard efficiency of the cycle.
For air, assume $n = 1.4$, $C_p = 1.005$ kJ/kg K, $C_v = 0.718$ kJ/kg K.
Assume a clearance volume of 10% of the total cylinder volume.

[Answer: 57.5%]

3.9 A Stirling engine uses air as the working fluid, operating with an isothermal compression at a temperature $T_C = 20$ °C and an isothermal expansion process at temperature $T_h = 200$ °C. It has a maximum volume of 600 cm³ and a minimum volume of 550 cm³. The lowest pressure in the cycle is 1 bar.
Estimate the cycle efficiency (%) and the net power output (W) when the engine speed is 50 cycles/s.
For air, take $R = 287$ J/kg K and $C_v = 718$ J/kg K.

[Answers: 38%, 160 W]

3.10 A diesel engine with a compression ratio of 17:1 is to operate on biodiesel fuel. If the ambient temperature is 27 °C and the maximum temperature is 1227 °C, calculate:
(a) The temperatures at the end of the compression and expansion processes.
(b) The net work output.
(c) The thermal efficiency of the cycle.
For air, assume $n = 1.4$, $C_p = 1.005$ kJ/kg K, $C_v = 0.718$ kJ/kg K.

[Answers: 931.7 K, 584 K, 296 kJ/kg, 64%]

4

Combustion

4.1 Overview

The overwhelming majority of both static and mobile power-generation systems in the world today rely on the burning of fossil fuels, i.e. a chemical reaction between a hydrocarbon compound and air that releases heat.

The energetic gases evolved may be used directly, as in a reciprocating engine or gas turbine, or they may give up their heat in energizing an intermediate working fluid like water/steam. In either case, the gases, which may prove to be environmentally harmful, are discharged to the atmosphere at the end of their usefulness.

This chapter will detail the combustion process from reactant species to product evolution in terms of mass and energy conservation. It will also introduce some of the more practical aspects of combustion and fuel use, for example, excess air, equivalence ratios and calorific values.

Learning Outcomes

- To understand the basic units of combustion chemistry.
- To be able to carry out simple chemical balances on combustion equations.
- To be introduced to basic combustion terminology related to reactant fuel–air mixtures.
- To be able to apply the first law of thermodynamics and conduct energy balances associated with the combustion process.
- To determine the resulting theoretical flame temperatures in a combustion process.
- To be introduced to the generation and removal of sulphur and nitrogen oxide emissions.

4.2 Mass and Matter

At a fundamental level, matter has been classified according to its atomic structure into *elements*, for example, carbon, hydrogen, argon. Elements are denoted by symbols or abbreviations of one or two letters, for example, C, H and Ar for carbon, hydrogen and argon, respectively.

Conventional and Alternative Power Generation: Thermodynamics, Mitigation and Sustainability, First Edition. Neil Packer and Tarik Al-Shemmeri.
© 2018 John Wiley & Sons Ltd. Published 2018 by John Wiley & Sons Ltd.

An element is a single homogenous substance having a clearly identifiable atomic structure that cannot be further subdivided by chemical means.

Matter may comprise a single element or a combination of elements.

Chemical combinations of atoms of the same element or atoms of different elements are called *molecules.* Molecules of any specific combination have a readily identifiable geometric structure characteristic of the combination and reflecting the inter-atomic relationship.

The bulk aggregation of a specific molecule is usually termed a *compound.*

Molecules are represented by formulae that consist of the element symbols and perhaps subscripts, for example, N_2 (nitrogen gas), SO_2 (sulphur dioxide), H_2O (water) etc. The subscripts denote the numbers of atoms of each element present. Note that if only one atom of an element is present, the subscript is omitted; for example, a molecule of carbon dioxide (CO_2) comprises one atom of carbon and two atoms of oxygen.

4.2.1 Chemical Quantities

When evaluating the magnitude of matter from a chemical point of view, the unit used is the *mole* or *kilomole*. A mole is defined as that amount of substance containing 6.0223×10^{23} molecules or atoms. This number is termed *Avogadro's number* or *Avogadro's constant.*

4.2.2 Chemical Reactions

At the most fundamental level, chemical reactions are the result of the migration of electrons and protons.

Chemical reactions describe, on a molecular scale, the combination of simple substances to form new ones and also the breakdown of complex substances into more simple ones.

In chemical reactions, substances are represented by their chemical symbols, with coefficients indicating the number of moles of substance present. For example:

$$H_2 + 0.5O_2 \rightarrow H_2O$$

In the above equation, one mole of hydrogen (H_2) combines with 0.5 moles of oxygen (O_2) to form one mole of water.

Note that coefficients having a value of 1 are omitted by convention.

The substances initially present in a reaction are termed the *reactants*. The substances resulting from a reaction are termed the *products*.

4.2.3 Physical Quantities

On a more macroscopic or everyday scale, the unit of mass is the *kilogram.*

There are many applications where a conversion between these two units, kilograms and kilomoles, is required and this is facilitated by a substance property called the *molar mass* (M, kg/kmol).

The *approximate* elemental molar masses of primary interest in this chapter are shown in Table 4.1.

Table 4.1 Table of common elements associated with combustion.

Element	Chemical symbol	Approximate molar mass (kg/kmol)
Carbon	C	12
Hydrogen	H	1
Oxygen	O	16
Nitrogen	N	14
Sulphur	S	32

(Note: Units of molar mass are sometimes quoted in a multiple-free ratio of g/mol).

The molar mass of molecules, or the *molecular mass*, can be found by simple accountancy. If a molecule contains i different elements, then the resulting molar mass can be calculated by summing the product of the number of atoms of each element with their attendant molar mass:

$$M_{molecule} - \sum \left[\text{number of atoms of element}_{(i)} \times M_{(i)}\right] \tag{4.1}$$

The mean molar mass of a mixture (M_{mix}) can be found with knowledge of component *molar fractions*.

The molar fraction (y_i) is simply the ratio of the number of moles (n_i) of the component of interest (i) to the total number of moles in the mixture (n_{tot}), i.e.:

$$y_i = n_i/n_{tot} \tag{4.2}$$

If a mixture contains i different components, then the resulting *molar mass of the mixture* can be calculated by summing the product of each component molar mass (M_i) with its attendant molar fraction (%) and dividing by a hundred; thus:

$$M_{mix} = \left(\sum(\text{molar fraction} \times M_i)\right)/100 \tag{4.3}$$

The relationship between the number of kilomoles of a substance (n), its mass (m, kg) and molar mass (M, kg/kmol) is given by:

$$m = nM \tag{4.4}$$

This relationship is extremely useful, as it provides for a conversion from chemical to physical units.

4.3 Balancing Chemical Equations

The principles of mass conservation also have an application to chemical reactions. Chemical reactions must be manipulated or *balanced* until the law of mass conservation is obeyed. To obtain a balanced equation, identify the products and reactants and set them out as described earlier. The following steps may then be employed:

Step 1: Set the coefficient (number of moles) of the first reactant to 1 and the coefficients of all the other substances (reactants and products) to unknowns a, b, c etc.

Step 2: Account for the number of atoms of each element in turn by multiplying the coefficient of that element by its subscript and equate between reactant and product. Solve the balance for each unknown coefficient, a, b, etc.; i.e. the number of moles of each substance.

Step 3: Rewrite the equation using the coefficient results from the balance.

Step 4: Check the result by using molar mass data to convert from moles to mass for each substance.

If the equation is balanced correctly, the total mass of reactants will equal the total mass of products.

[Note that numbers of atoms and mass must always balance between reactants and products, however, numbers of moles may not.]

4.3.1 Combustion Equations

Typically, in a combustion process we have:

Fuel + oxidizer $(+$ ignition$)$ → Products of combustion

(reactants)   **(products)**

When the fuel is a *hydrocarbon*, for example, methane (CH_4), ethane (C_2H_6), octane (C_8H_{18}) then:

For the carbon $C + O_2 \rightarrow CO_2$
For the hydrogen $H_2 + 0.5O_2 \rightarrow H_2O$

For combustion of a hydrocarbon fuel with air, a balanced equation can be obtained by using the method outlined previously, with both reactions being considered simultaneously.

Combustion reactions usually take place with *ambient air* that consists of 21% O_2 and 79% N_2 by volume. So every mole of oxygen brings with it (79/21) 3.76 moles of N_2. *This must be accounted for when balancing combustion equations.*

Commonly, N_2 is assumed not to undergo any chemical change in simple combustion analysis. In reality, oxides of nitrogen are formed if the temperature is high enough.

Again, in simple analyses, the effects of other atmospheric components present in small concentrations, such as inert gases, water vapour etc., are ignored.

If the products of combustion comprise CO_2, H_2O, N_2, as above (and perhaps O_2, as will be shown later) then we have *complete combustion*.

Incomplete or *partial combustion* is associated with insufficient air supply or poor mixing and results in products that may contain H_2, CO, C, OH and N_2.

Any hydrogen or carbon in the products must be regarded as *wasted fuel*, as it could potentially be reacted with oxygen to release its energy.

4.4 Combustion Terminology

4.4.1 Oxidizer Provision

The minimum amount of air that supplies sufficient O_2 for complete combustion in the previous examples is termed the *100% theoretical or stoichiometric* (*perfectly chemically proportioned*) quantity of air.

The term *air–fuel ratio* (*A/F*) is defined as the ratio of the mass of air to the mass of fuel in a combustion process, i.e.

$$(A/F) = m_{air}/m_{fuel} = \sum(nM_{air})/\sum(nM_{fuel}) \qquad (4.5)$$

$$= \frac{(n_{air})(1 + 3.76)(29)}{(\text{number of carbon atoms}) (12) + (\text{number of hydrogen atoms}) (1)}$$

Due to imperfect mixing between fuel and air, it is common practice to supply an amount of air *in excess* of the stoichiometric amount to achieve complete combustion.

To describe this amount, the terms *% excess air* and *% stoichiometric air* (or *% theoretical air*) are in common usage.

The % excess air is commonly quoted in terms of the air–fuel ratio, thus:

$$\% \text{ excess air} = \left(\frac{(A/F)_{actual} - (A/F)_{stoichiometric}}{(A/F)_{stoichiometric}} \right) \times 100 \qquad (4.6)$$

and

$$\% \text{ stoichiometric air} = 100\% + \% \text{ excess air} \qquad (4.7)$$

$$= (n_{actual}/n_{stoichiometric})$$

where *n* = number of moles of air present.

Note that if excess air is employed, the unused oxygen and its attendant nitrogen will appear in the products. This must be accounted for when carrying out mass and energy balances.

Another term used to describe the state of the fuel–air mixture is the *equivalence ratio* (*ϕ*), defined thus:

$$\text{Equivalence ratio, } \phi = \frac{(A/F)_{stoichiometric}}{(A/F)_{actual}} = \frac{(F/A)_{actual}}{(F/A)_{stoichiometric}}$$

$$= \frac{100\%}{\% \text{ stoichiometric air}} \qquad (4.8)$$

ϕ > 1 indicates a (*fuel*) rich mixture.
ϕ = 1 indicates stoichiometric proportions.
ϕ < 1 indicates a (*fuel*) lean mixture.

4.4.2 Combustion Product Analyses

Sampling and analysis of product gases is commonly employed to determine the state of fuel–air combustion mixtures.

Analyzers are usually *dry-gas* based, i.e. the water vapour is removed before exposure to the analyzer's sensors. However, the balancing technique detailed above is still applicable providing that the water is reinstated into the product side of the combustion equation.

4.4.3 Fuel mixtures

The above discussion is limited to the combustion of a *pure* fuel, i.e. one comprising only a single species of hydrocarbon. In reality, some fuels can be complex mixtures, for example, natural gas is a mixture of methane, ethane and heavier hydrocarbons as well as some nitrogen and carbon dioxide. Coal and biomass-sourced fuels can be even more complex. Nevertheless, the combustion equation-balancing technique can still be applied to first approximation.

4.5 Energy Changes During Combustion

The *standard enthalpy of formation* ($\bar{h}_f^\circ$) is the enthalpy of a substance at a specified reference state (usually 298.15 K, 1 atm, denoted by the superscript '°') due to its chemical composition and is associated with the breaking and making of chemical bonds.

Compounds with a negative enthalpy of formation release energy when they are formed in an *exothermic* reaction. Compounds with a positive enthalpy of formation require energy for their creation in an *endothermic* reaction.

Note that, *by convention*, the enthalpy of formation of some stable substances has been traditionally set to zero, i.e. ($\bar{h}_f^\circ = 0$).

The heat or *enthalpy of vaporization* ($\bar{h}_{fg}$) is the heat required to completely vaporize a unit quantity of saturated liquid. Combustion calculations are generally performed with fuels in a gaseous (g) state. However, many fuels are supplied in a liquid form. Liquid fuels must be vaporized prior to combustion and the enthalpy of vaporization quantifies this phase change energy requirement. This must be accounted for when performing energy balances associated with combustion involving liquid fuels.

When reviewing data tables, it will be noted that there are two enthalpy of formation entries for water. The applicable value depends on whether the water exists in a liquid (l) or in a gaseous (g) state. The difference in numerical value is accounted for by the enthalpy of vaporization value for water.

The heat or *enthalpy of reaction* ($\bar{h}_r$) is the difference between the enthalpy of the combustion products at a specified state (i.e. temperature, pressure) and the enthalpy of the reactants *at the same state* for a complete reaction.

In some texts and handbooks, the term *heat of combustion* is used. This parameter is the enthalpy of reaction for a combustion process (and is sometimes quoted with a sign change).

Values of $\bar{h}_f^\circ$ and $\bar{h}_r$ for some common hydrocarbons and other compounds are presented on a molar-independent basis in Table B.1 in Appendix B. Note also that thermodynamic properties quoted per kmol are identified by the addition of the *bar* accent.

Molar-independent *sensible enthalpy* values ($\bar{h} - \bar{h}^\circ$) for gases at 1 atm and temperatures *relative to the standard state* are presented in Table B.2 in Appendix B.

At any given state, the *total enthalpy* ($\bar{h}_{tot}$) of a chemical component is the sum of the enthalpy of formation and sensible enthalpy of the component, i.e.:

$$\bar{h}_{tot} = \bar{h}_f^\circ + (\bar{h} - \bar{h}^\circ) \tag{4.9}$$

Another term commonly used to describe the energetic content of commercial fuel is the *heating or calorific value* (MJ/kg or MJ/m³). This value will vary according to whether any water vapour in the products is condensed, thus releasing its latent heat of condensation.

If the water vapour is condensed, the total energy release per unit quantity of fuel is known as the *higher heating value (HHV)* or *gross calorific value (GCV)*. If not, the energy release is known as the *lower heating value (LHV)* or *net calorific value (NCV)*.

Heating values can be estimated by a simple conversion of the heat of combustion value from molar to mass terms.

The higher heating values of some common hydrocarbons are also given in Table B.1 in Appendix B. Specified pressure and temperature conditions should be associated with any published values.

4.6 First Law of Thermodynamics Applied to Combustion

4.6.1 Steady-flow Systems (SFEE) [Applicable to Boilers, Furnaces]

Applying the first law of thermodynamics to a combustion control volume and neglecting any potential or kinetic energy changes, the steady flow chemical reaction can be written as:

$$\dot{Q} - \dot{W} = \sum_{\text{products}} \dot{n}_i \left[\bar{h}_f^o + (\bar{h} - \bar{h}^o) \right]_i - \sum_{\text{reactants}} \dot{n}_i \left[\bar{h}_f^o + (\bar{h} - \bar{h}^o) \right]_i \tag{4.10}$$

This form requires knowledge of the molar flow rates ($\dot{n}$) through the volume.

However, it is more common in combustion studies to express this relationship in *per mole of fuel* terms and to divide the above equation through by the molar flow rate of fuel. (Note that the LHS terms lose their 'dot' notation as a consequence.)

Assuming zero work exchange (and zero heat input), this becomes:

$$Q = \sum_{\text{products}} n_i \left[\bar{h}_f^o + (\bar{h} - \bar{h}^o) \right]_i - \sum_{\text{reactants}} n_i \left[\bar{h}_f^o + (\bar{h} - \bar{h}^o) \right]_i \tag{4.11}$$

If the process is *isothermal*, the sensible enthalpy terms disappear and this further reduces to:

$$Q = \sum_{\text{products}} n_i \left[\bar{h}_f^o \right]_i - \sum_{\text{reactants}} n_i \left[\bar{h}_f^o \right]_i \tag{4.12}$$

4.6.2 Closed Systems (NFEE) [Applicable to Engines]

Again, neglecting any kinetic or potential energy changes:

$$Q - W = \sum_{\text{products}} n_i \left[\bar{u}_f^o + (\bar{u} - \bar{u}^o) \right]_i - \sum_{\text{reactants}} n_i \left[\bar{u}_f^o + (\bar{u} - \bar{u}^o) \right]_i \tag{4.13}$$

Internal energy tables are available, but to avoid using another property, advantage may be taken of the enthalpy/internal energy/flow work relationship, thus:

$$Q - W = \sum_{\text{products}} n_i \left[\bar{h}_f^o + (\bar{h} - \bar{h}^o) - \mathrm{P}v \right]_i - \sum_{\text{reactants}} n_i \left[\bar{h}_f^o + (\bar{h} - \bar{h}^o) - \mathrm{P}v \right]_i$$

$$= \sum_{\text{products}} n_i \left[\bar{h}_f^o + (\bar{h} - \bar{h}^o) - R_o T \right]_i$$

$$- \sum_{\text{reactants}} n_i \left[\bar{h}_f^o + (\bar{h} - \bar{h}^o) - R_o T \right]_i \tag{4.14}$$

4.6.3 Flame Temperature

The *adiabatic flame temperature* is the ideal maximum temperature of the resulting combustion flame assuming no work, heat loss or changes in kinetic and potential energy, and is calculated from knowledge of the fuel energy content and the heat capacities of the combustion products.

For a *constant-pressure* process, such as a gas turbine combustor or the burner of a furnace, these conditions reduce the first law of thermodynamics to:

$$\sum_{\text{reactants}} n_i \bar{h}_i = \sum_{\text{products}} n_i \bar{h}_i \tag{4.15}$$

For a *constant-volume* process, such as part of a theoretical engine cycle, the first law becomes:

$$\sum_{\text{reactants}} \left[(n_i \bar{h}_i) - (n_i R_o T_i) \right] = \sum_{\text{products}} \left[(n_i \bar{h}_i) - (n_i R_o T_i) \right] \tag{4.16}$$

Since the temperature of the products is unknown at the outset, adiabatic flame temperature calculation involves an iterative process, as follows:

Perform a balance for the combustion equation accounting for any prevailing conditions, for example, excess air etc.

Guess a temperature for the products.

Using enthalpy tables, perform a reactant–product energy balance at the estimated temperature.

For example, in the case of a constant-pressure process:
If $\sum_{\text{reactants}} n_i \bar{h}_i = \sum_{\text{products}} n_i \bar{h}_i$ then the guess is correct.
In the absence of equality, guess a new product temperature using the magnitude and sign of the result of the previous guess as a guide. Re-attempt a balance.

If still unlucky, interpolate over a small interval, say 200 K, to solve for a reactant–product balance and the resulting adiabatic flame temperature.

(As an alternative, plot the results on an enthalpy–temperature basis and read off.)

Actual flame temperatures are always less than this theoretical value due to heat loss from the flame by convection/radiation and unwanted *dissociation* (reverse chemical reactions) of the combustion products that consume energy.

4.7 Oxidation of Nitrogen and Sulphur

Hydrocarbon fossil fuels are commonly more than just hydrogen and carbon in composition and can comprise a range of other elements and compounds, albeit in relatively small amounts. For example, sulphur can be found in solid and liquid fuels and, in association with hydrogen, in gaseous fuels. Sulphur will burn with oxygen in a combustion process but, unfortunately, the resulting products of combustion are unwelcome, being both poisonous and able to go on to become corrosive.

Fuels may also contain nitrogen that can, under certain circumstances, also play an unwanted part in a combustion process. This occurrence is greatly exacerbated by the presence of nitrogen in the combustion oxidant gas that is usually atmospheric air. Again, the resulting products of combustion are able to go on to become a noxious emission.

4.7.1   Nitrogen and Sulphur

Elemental nitrogen is a relatively inert and harmless non-metal. It was first identified as an element in 1772 by British chemist Daniel Rutherford. It has a relative atomic mass of approximately 14 and boils at $-196\,°C$, so at normal temperature and pressure it is a gas.

Air in the lower atmosphere comprises approximately 79% nitrogen by volume and 76% by mass. Nitrogen makes up about 48% of all dissolved oceanic gas.

In combination with other elements, it is essential for life and it is to be found in many bio-compounds, including DNA. It can be fixed directly from the air by some soil bacteria. When converted into nitrates it can be made available to plants.

It is not truly inert, as, under the right conditions, it will react with hydrogen, oxygen and some metals.

Nitrogen reacts with hydrogen to form ammonia (NH_3). Its compounds are essential in the manufacture of many products including fertilizer and explosives.

It is also used in processes requiring an operating environment with a low reactivity, and in its liquid form it is used as a coolant.

Elemental sulphur is a bright yellow non-metal that can exist in three forms – as a powder, in crystalline form or as a soft solid. It was known to the ancient Greeks and Egyptians. French scientists Louis-Josef Gay-Lussac and Louis-Jacques Thénard first identified it as an element in 1809.

It has a relative atomic mass of approximately 32. It melts at $115\,°C$ and boils at $445\,°C$.

Its natural occurrence is closely associated with volcanic activity and some fossil fuel reserves. It is malodorous and hazardous to health when in combination with hydrogen or when burnt.

Again, it is important in biochemistry as a component of some amino acids.

Its compounds are commonly used in bleaches and preservatives and are used in the production of sulphuric acid (H_2SO_4), which is a component that underpins a myriad of manufacturing processes.

4.7.2   Formation of Nitrogen Oxides (NO_x)

The reduction (i.e. the addition of hydrogen) of nitrogen produces ammonia (NH_3):

$$N_2 + 3H_2 \leftrightarrows 2NH_3$$

Oxidation of nitrogen produces, progressively, nitric oxide (NO) then nitrogen dioxide (NO_2).

In short:

$$0.5N_2 + 0.5O_2 \leftrightarrows NO$$

$$NO + 0.5O_2 \leftrightarrows NO_2$$

At high levels of concentration, NO_2 is a lung irritant, causing inflammation of the airways and bronchitis-like symptoms. The World Health Organization's time-based concentration limits for exposure to the gas (2016) are:

- $40\,\mu g/m^3$ – annual mean
- $200\,\mu g/m^3$ – 1-hour mean

NO_x in combination with hydrocarbon compounds (HC) also plays a part in the formation of low-level ozone (O_3) which, again, is a lung irritant.

$$NO_x + HC + \text{sunlight} + O_2 \rightarrow O_3 + \text{photochemical smog}$$

There are three main combustion-sourced mechanisms of NO_x formation:

- *Thermal NO_x* formed by the combination of nitrogen and oxygen in high-temperature (>1800 K) combustion systems over a range of equivalence ratios.
  The commonly used model of thermal NO_x formation is called the (*extended*) *Zeldovich mechanism*.
  This assumes that O radicals attack N_2 molecules, thus:

  $$N_2 + O \leftrightarrows NO + N$$

  The resulting N radicals can form NO by the following reaction:

  $$N + O_2 \leftrightarrows NO + O$$

  Nitrogen can also combine with hydroxyl radicals (OH), thus:

  $$OH + N \leftrightarrows NO + H$$

  Its formation is slow compared to the attendant fuel–oxygen combustion reactions and so its generation is generally limited to the exhaust gas environment.

- *Prompt NO_x* from reaction of fuel-derived radicals with N_2, ultimately leading to NO. The commonly used model of NO_x formation, in this case, is called the *Fenimore mechanism* and is more prevalent in fuel-rich mixtures.
  As suggested by its denomination, its formation is faster than thermal NO_x reactions and its generation can take place in the flame zone environment.
  The mechanism assumes the presence of hydrocarbon radicals (CH, C, H) to initiate the process and generate nitrogen (N), thus:

  $$CH + N_2 \leftrightarrows HCN + N$$

  $$C + N_2 \leftrightarrows CN + N$$

  Alternative routes to NO_x generation are possible.
  One of the most commonly quoted paths utilizing hydrogen cyanide (HCN) is:

  $$HCN + O \leftrightarrows NCO + H$$

  $$NCO + H \leftrightarrows NH + CO$$

  $$NH + H \leftrightarrows N + H_2$$

  $$N + OH \leftrightarrows NO + H$$

- *N_2O intermediate NO_x* is generated in combustion systems with equivalence ratios of less than 1, i.e. fuel-lean mixtures, and requires the presence of an unchanged, facilitating chemical species (M) and hydrogen and oxygen radicals, thus:

  $$O + N_2 + M \leftrightarrows N_2O + M$$

  $$N_2O + H \leftrightarrows NH + NO$$

  $$N_2O + O \leftrightarrows NO + NO$$

  N_2O and prompt NO_x are only weakly dependent on temperature.

4.7.3 NO$_x$ Control

Two approaches to NO$_x$ control are possible: combustion process modification or post flame/exhaust gas treatment.

4.7.3.1 Modify the Combustion Process

Several possibilities present themselves:

- Eliminate the atmospheric nitrogen by the use of pure oxygen for combustion. This is a very expensive solution.
- Manipulate the air–fuel ratio.
- Operating on a fuel-rich basis reduces the amount of available oxygen and related nitrogen. Operating on a fuel-lean basis, i.e. providing high excess air, increases the amount of nitrogen but reduces the resulting flame temperature.
- Recirculation of the flue gases to the combustion chamber to reduce peak flame temperature. NO$_x$ reductions of 70–80% are possible.
- Water injection into the combustion chamber, hence reducing the temperature due to the use of energy in vaporizing the water.
- Air-staged (rich–lean) combustion or low-NO$_x$ burners producing a twin-peaked temperature–time characteristic.
- In the first stage, the air–fuel ratio is such that all the O$_2$ is used up, leaving none for NO$_x$ formation.
- In the second stage, due to the heat released in the first stage, the temperature is limited below that for large-scale NO$_x$ formation.
- Highly effective mixing and/or inter-stage cooling is required. With this method, NO$_x$ reduction of 40–70% is possible; however, HC and CO emissions may increase in the exhaust gases.
- Re-burning, which is another staged combustion strategy employing the staging of fuel and air.

4.7.3.2 Post-flame Treatment

Most post-flame treatments either add a reducing agent to the combustion gas stream to take oxygen away from NO or pass the NO$_x$ over a catalyst.

For power plants and large furnaces, *ammonia* (or urea) is commonly added as a reducing agent.

$$6NO + 4NH_3 \leftrightarrows 5N_2 + 6H_2O$$

and

$$6NO_2 + 8NH_3 \leftrightarrows 7N_2 + 12H_2O$$

Note the above reducing reactions are only effective at temperatures of between 850 and 1000 °C. Above 1000 °C the following reaction may occur, *increasing* NO formation:

$$4NH_3 + 5O_2 \leftrightarrows 4NO + 6H_2O$$

Below 850 °C the conversion to N$_2$ and water vapour is unacceptably slow.
NO$_x$ reductions of 60–80% are possible.
In a modern car engine, post-flame treatment is carried out using a catalytic converter:

$$2NO + 2CO \leftrightarrows N_2 + 2CO_2$$

However, control over the NO/CO ratio is important for good conversion.

4.7.4   Formation of Sulphur Oxides (SO$_x$)

Oxides of sulphur are air pollutants.

Elemental sulphur will react with oxygen in the air to produce sulphur dioxide (SO$_2$):

$$S + O_2 \leftrightarrows SO_2$$

The rate of the reaction increases with increasing temperature.

Sulphur dioxide (SO$_2$) is an air pollutant causing irritation and inflammation to the respiratory tract and eyes at high levels of concentration. The World Health Organization's time-based concentration limits for exposure to the gas (2016) are:

- 20 µg/m^3 – 24-hour mean
- 500 µg/m^3 – 10-minute mean.

Sulphur in contact with hydrogen can be reduced to hydrogen sulphide (H$_2$S), also regarded as an air pollutant.

4.7.5   SO$_x$ Control

Two fuel-related sources of sulphur emissions will be considered here.

4.7.5.1   Flue Gas Sulphur Compounds from Fossil-fuel Consumption

Sulphur is commonly found in coal and oil, so SO$_2$ in low (<0.5%) concentrations will be a minor component of their combustion gases. The total volume of the emission on a global scale is, however, very large.

Scrubbing systems for this problem are called *flue gas desulphurization* (FGD) units.

For this application, three possible removal arrangements are commonly in use: gas bubblers, spray chambers and packed chambers.

In a simple bubbler, the waste gas is forced, under pressure, through a perforated pipe submerged in the scrubbing liquid. With small bubbles, good contact will result. However, large gas-side pressure drops may occur in tall towers.

In a spray chamber, the gas flows up an open chamber through a shower of descending scrubbing liquid released from spray nozzles. Gas–liquid contact is inferior to a bubbler but lower pressure drops prevail.

A packed column bed (see Figure 4.1) is a modified spray chamber with the column interior filled with some kind of solid medium on which the liquid can form a thin film, giving a large surface area to facilitate mass transfer. The medium or bed can comprise engineered ceramic, plastic or metal shapes or even natural materials such as rocks/gravel. This type of arrangement provides the best mass transfer per unit pressure drop performance.

In this application, the SO$_2$ is ultimately converted to calcium sulphate (CaSO$_4$) by using limestone (CaCO$_3$) slurry as the liquid in a wet scrubber. This facilitates its capture and removal. The basic chemical reaction is:

$$2CaCO_3 + 2SO_2 + O_2 \leftrightarrows 2CaSO_4 + 2CO_2$$

A typical arrangement for calcium sulphate removal and scrubbing liquid reuse is shown in Figure 4.2. The principal components of the system are the scrubber, holding tank and settling chamber.

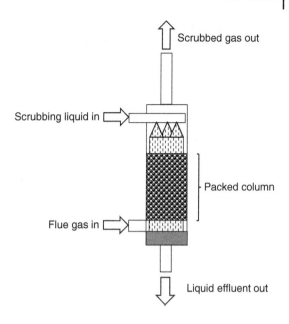

Figure 4.1 Packed column SO$_2$ scrubbing tower arrangement.

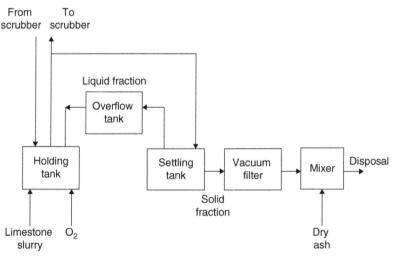

Figure 4.2 Typical flue gas desulphurization (FGD) waste recovery system.

The liquid exiting the scrubber is transferred to a holding tank where finely ground limestone and oxygen are added.

The resulting slurry is recirculated from the holding tank to the scrubber liquid inlet.

A fraction of the holding tank recirculation is transferred to a settling chamber to facilitate solid removal.

The outflow of the settling chamber has its water content reduced by vacuum filtration and is then often mixed with dry flyash (particulate matter from the combustion process) for easier handling before going to landfill. The liquid fraction overflow of the settling chamber is returned to the holding tank.

Liquid circulation rates are very large in these types of scrubbers, with the slurry spending only a few seconds in the tower and up to ten minutes in the holding tank, where most of the reactions take place.

In modern plant, it is becoming less common for an FGD's output to end up in landfill. Many plants are able to sell their calcium sulphate waste on to the construction supply industry for the manufacture of gypsum-based plasters.

Limestone scrubbers have associated with them many operational considerations. For example:

- Scrubbing cools the flue gas and so there is a requirement to reheat the cleaned flue gas to maintain plume buoyancy for later dispersal. Reheating is also required to raise the temperature to above the acid dew point, preventing any acid condensation and corrosion downstream of the scrubber.
- As indicated earlier, the basic removal of limestone reaction generates carbon dioxide.
- Solid deposition either via entrainment of slurry droplets in the exhaust system or plugging and scaling of FGD-associated fluid flow systems, e.g. pumps, valves etc., is a problem.

This list is not exhaustive.

4.7.5.2 Sulphur Compounds from Petroleum and Natural Gas Streams

Some natural gas streams contain greater than 1% by weight of hydrogen sulphide (H_2S), which must be removed to make the gas acceptable for domestic use. Gases with high H_2S concentrations are often described as *sour* and their removal as *sweetening*.

Absorber/stripper arrangements utilizing amine solutions are commonly employed for its removal. Sulphur recovered by this process has a market value, making a significant contribution to global supply. This technique is also used in some proposed CO_2 removal systems. More detail on gas absorption is supplied in Chapter 6.

4.7.6 Acid Rain

The natural acidity of freshwater sources such as lakes, rivers and streams is dependent on several factors including rainfall as well as the local geology and ecosystem.

For example, as a result of a reaction with atmospheric carbon dioxide, carbonic acid (H_2CO_3) can be formed in naturally occurring rainwater.

$$H_2O + CO_2 \leftrightarrows H_2CO_3$$

$$H_2CO_3 \leftrightarrows HCO_3^- + H^+$$

Rainfall pH values of around 5.6 are possible.

The pH of freshwater can also be altered by the composition of run-off and leachate entering its volume after passage through/over surrounding rocks and soils.

However, as a consequence of nitrogen and sulphur oxide reactions with atmospheric water, more extreme acid deposition, more commonly referred to as acid rain, can result.

In the presence of water vapour in the atmosphere, NO_2 can react to form nitric acid (HNO_3):

$$3NO_2 + H_2O \leftrightarrows 2HNO_3 + NO$$

Oxidation of sulphur dioxide produces sulphur trioxide (SO_3):

$$SO_2 + 0.5O_2 \leftrightarrows SO_3$$

Upon contact of SO_3 with atmospheric water, sulphuric acid (H_2SO_4) will be formed:

$$SO_3 + H_2O \leftrightarrows H_2SO_4$$

As a result of the above reactions, localized and trans-national freshwater lake pH levels can be as low as 3.5 in some areas. The impact is exacerbated by the fact that some elemental metals become soluble at elevated pH values and so can be dissolved out of rocks and soils then washed into feeder streams and rivers, where they can affect the developmental biology of a number of species.

Sulphur-based acid rain can be particularly damaging to buildings constructed from carbonate-based materials that react with the pollutant, replacing the carbonates with less-cohesive sulphate compounds (gypsum) that have a tendency to blacken and flake from the building surface.

The conversion process from air pollutant gas to water pollutant is not instantaneous and, in some cases, oxide gases can be carried by the prevailing wind far from their source before acid deposition occurs. For example, much of the historical acid deposition experienced in Scandinavian forests is thought to have originated In other, predominantly coal-burning areas of northern Europe such as Germany and the UK. A similar set of circumstances prevails between the coal-burning plants of the north-eastern states of the USA and Canada.

4.8 Worked Examples

Worked Example 4.1 – Determination of molar mass of a molecule
Determine the molar mass (kg/kmol) of a carbon dioxide (CO_2) molecule.

Solution
Find M_{CO_2}.

Molar mass of carbon (C): 12 kg/kmol
Molar mass of oxygen (O): 16 kg/kmol.

Since carbon dioxide contains one carbon and two oxygen atoms:

$$M_{CO_2} = [(1 \times M_C) + (2 \times M_O)]$$
$$= [(1 \times 12) + (2 \times 16)]$$
$$= 44 \, kg/kmol.$$

Worked Example 4.2 – Determination of molar mass of a gas mixture
Consider atmospheric air to consist of 79% nitrogen (N_2) and 21% oxygen (O_2) by molar fraction. Determine the mean molar mass (kg/kmol) for the mixture.

Solution
Given: Molar fraction (%) N_2, O_2

Find $M_{mean,air}$.

The molar masses (kg/kmol) of each component are:

$$M_{N_2} = 2 \times M_N = 2 \times 14 = 28 \text{ kg/kmol}$$

$$M_{O_2} = 2 \times M_O = 2 \times 16 = 32 \text{ kg/kmol}$$

The average molar mass for *air* is:

$$M_{air} = \left(\sum (\text{molar fraction} \times M_i) \right) / 100$$

$$= [(\%N_2 \times 28) + (\%O_2 \times 32)]/100$$

$$= [(79 \times 28) + (21 \times 32)]/100$$

$$= 28.84 \text{ kg/kmol}$$

[This will be approximated by 29 kg/kmol in the following calculations.]

Worked Example 4.3 – A moles-to-mass conversion

What is the mass (kg) of 3 kmoles of carbon dioxide?

Take the molar mass of carbon dioxide (CO_2) to be 44 kg/kmol.

Solution

Given: $n = 3$ kmol, $M = 44$ kg/kmol.
Find m.

$$m = nM$$

$$= 3 \times 44$$

$$= 132 \text{ kg}$$

Worked Example 4.4 – A mole/mass balance for a simple fuel

Using the molar mass data below, perform a mass balance for the reaction of a hydrogen molecule (H_2) with an oxygen molecule (O_2) to form water (H_2O).

Molar mass data (M, kg/kmol)
Hydrogen (H): 1
Oxygen (O): 16
Water (H_2O): ($[2 \times 1] + 16$) $= 18$

Solution

Using the methodology outlined above:

Step 1: $H_2 + aO_2 \rightarrow bH_2O$
Step 2: Perform a balance

Element	Number of atoms in reactants	Number of atoms in products	Balance result
H	2(1)	2b	$b = 2/2 = 1$
O	2a	b	$a = b/2 = 1/2 = 0.5$

Step 3: $H_2 + 0.5O_2 \rightarrow H_2O$
Step 4: using $m = nM$, we have:

Reactants

Mass of hydrogen: $m_{H_2} = 1 \times (2 \times 1) = 2$ kg
Mass of oxygen: $m_{O_2} = 0.5 \times (2 \times 16) = 16$ kg
Thus, the total mass of reactants is 18 kg.

Products

Mass of water: $m_{H_2O} = 1 \times [(2 \times 1) + 16] = 18\,$kg

Since the mass of products is equal to the mass of reactants, the mass of product in the equation arrived at in Step 3 is correct.

Worked Example 4.5 – A mole/mass balance for a hydrocarbon fuel
Determine a balanced chemical equation for the combustion of ethane (C_2H_6) with air.

Solution
Note here that combustion equations are usually balanced with the number of moles of fuel equal to unity.

$$C_2H_6 + a(O_2 + 3.76N_2) \rightarrow bCO_2 + cH_2O + dN_2$$

Balancing between reactants and products:

Element	Number of atoms in reactants	Number of atoms in products	Balance result
C	2	b	$b = 2$
H	6	$2c$	$c = 6/2 = 3$
O	$2a$	$2b + c$	$a = [2(2) + 3]/2 = 3.5$
N	$2(3.76)(a)$	$2d$	$d = 2(3.76)(3.5)/2 = 13.16$

The balanced equation is, therefore:

$$C_2H_6 + 3.5(O_2 + 3.76N_2) \rightarrow 2CO_2 + 3H_2O + 13.16N_2$$

Worked Example 4.6 – A balance for non-stoichiometric conditions
Ethane (C_2H_6) is combusted with 90% of the air required to completely burn all of its carbon to carbon dioxide (CO_2), resulting in carbon monoxide (CO) in the combustion products. If the hydrogen in the fuel is completely burnt to water (H_2O), determine the balanced combustion equation.

Solution

From the previous example, the equation for the complete combustion of ethane is:

$$C_2H_6 + 3.5(O_2 + 3.76N_2) \rightarrow 2CO_2 + 3H_2O + 13.16N_2$$

With 90% of the supply air and carbon monoxide (CO) in the products:

$$C_2H_6 + (0.9)3.5(O_2 + 3.76N_2) \rightarrow bCO_2 + cH_2O + dN_2 + eCO$$

Balancing between reactants and products:

Element	Number of atoms in reactants	Number of atoms in products	Balance result
C	2	$b + e$	$b + e = 2$
H	6	$2c$	$c = 6/2 = 3$
O	$(0.9)(3.5)2$	$2b + c + e$	$2b + e = 3.3$
N	$(0.9)(3.5)(3.76)2$	$2d$	$d = 23.68/2 = 11.84$

Note that as the hydrogen is burnt completely, the moles of water vapour remain unchanged.

Using the balance results for carbon (C) and oxygen (O) as simultaneous equations:

$$b + e = 2$$
$$2b + e = 3.3$$
$$b = 1.33$$

and hence, $e = 0.67$.

The balanced equation is, therefore:

$$C_2H_6 + (0.9)3.5(O_2 + 3.76N_2) \rightarrow 1.33CO_2 + 3H_2O + 11.84N_2 + 0.67CO$$

Worked Example 4.7 – Use of the air–fuel ratio

A methane (CH_4) – air mixture has an A/F ratio of 20.

Determine the stoichiometric A/F ratio, the % excess air, the % stoichiometric air, the equivalence ratio and the molar fraction of CO_2 in the products.

Solution

$$CH_4 + a(O_2 + 3.76N_2) \rightarrow bCO_2 + cH_2O + dN_2$$

Balancing between reactants and products:

Element	Number of atoms in reactants	Number of atoms in products	Balance result
C	1	b	$b = 1$
H	4	$2c$	$c = 4/2 = 2$
O	$2a$	$2b + c$	$a = [2(1) + 2]/2 = 2$
N	$2(3.76)(a)$	$2d$	$d = 2(3.76)(2)/2 = 7.52$

The stoichiometric equation is:

$$CH_4 + 2(O_2 + 3.76N_2) \rightarrow CO_2 + 2H_2O + 7.52N_2$$

$$(A/F)_{sto} = m_{air}/m_{fuel} = \frac{(2 \times (1 + 3.76)) \times 29}{(1 \times 12) + (4 \times 1)} = 17.26$$

$$\% \text{ excess air} = \left(\frac{(A/F)_{actual} - (A/F)_{stoichiometric}}{(A/F)_{stoichiometric}} \right) \times 100$$

$$= \left(\frac{20 - 17.26}{17.26} \right) \times 100 = 15.87\%$$

$$\% \text{ stoichiometric air} = 100\% + \% \text{ excess air} = 100 + 15.87 = 115.87\%$$

$$\text{Equivalence ratio, } \phi = \frac{(A/F)_{stoichiometric}}{(A/F)_{actual}} = \frac{17.26}{20} = 0.863, \text{ i.e. fuel lean}$$

To calculate the molar fraction of CO_2 in the products, the actual combustion equation needs to be determined.

Revisiting the original combustion equation and allowing for the excess air in the actual case:

$$CH_4 + (1.1587)(2)(O_2 + 3.76N_2) \rightarrow CO_2 + 2H_2O + eO_2 + dN_2$$

The air supply (a) has increased by a factor equal to the % stoichiometric air/100, resulting in amount (e) of unused oxygen in the products. This will, of course, also affect the amount of nitrogen (d) in the products.

However, note that the number of moles of carbon dioxide (b) and water (c) remain the same, since the supply of excess air does not influence the fuel supply.

On a re-balance:

Element	Number of atoms in reactants	Number of atoms in products	Balance result
C	1	b	$b = 1$
H	4	$2c$	$c = 4/2 = 2$
O	(1.1587)(2)(a)	$2b + c + 2e$	$e = (4.6348 - 4)/2 = 0.317$
N	(1.1587)2(3.76)(2)	$2d$	$d = 17.427/2 = 8.713$

The actual equation is, therefore:

$$CH_4 + (1.1587)(2)(O_2 + 3.76N_2) \rightarrow CO_2 + 2H_2O + 0.317O_2 + 8.713N_2$$

The molar fraction of CO_2 in the products is given by:

$$\text{Molar fraction of } CO_2 \text{ in products, } y_{CO_2} = \frac{\text{Number of moles of } CO_2}{\text{Total moles of products}}$$

$$= [1/(1 + 2 + 0.317 + 8.713)] = 0.083 = 8.31\%$$

Worked Example 4.8 – Working with flue gas analyses

The results of a *dry gas* product volumetric analysis on the combustion of a methane–air mixture are as follows:

Carbon dioxide, CO_2: 10%
Carbon monoxide, CO: 1%
Oxygen, O_2: 5%
Nitrogen, N_2: 84%.

Determine the air–fuel ratio and the equivalence ratio of the combustion mixture.

Solution
In this case, the number of moles of fuel and air is unknown.

Let the combustion mixture contain (b) moles of fuel and a reinstated amount of product water (c).

Then, assuming 100 kmols of *dry* products,

$$bCH_4 + a(O_2 + 3.76N_2) \rightarrow 10CO_2 + 1CO + 5O_2 + 84N_2 + cH_2O$$

Balancing between reactants and products:

Element	Number of atoms in reactants	Number of atoms in products	Balance result
C	b	$10 + 1$	$b = 11$
H	$4b$	$2c$	$c = (4)11/2 = 22$
N	$2(3.76)(a)$	2×84	$a = 2 \times 84/(2 \times 3.76) = 22.34$

i.e. $11CH_4 + 22.34(O_2 + 3.76N_2) \rightarrow 10CO_2 + 1CO + 5O_2 + 84N_2 + 22H_2O$

Divide through by 11 to obtain an equation based on 1 mol of fuel:

$$CH_4 + 2.03(O_2 + 3.76N_2) \rightarrow 0.9CO_2 + 0.09CO + 0.45O_2 + 7.63N_2 + 2H_2O$$

The actual air–fuel ratio is given by:

$$(A/F) = m_{air}/m_{fuel} = \sum(nM_{air})/\sum(nM_{fuel})$$

$$= \frac{(2.03)(1 + 3.76)(29)}{(1)(12) + (4)(1)} = 17.51$$

[Remembering from the previous example that the stoichiometric air–fuel ratio for methane was 17.26.]

$$\text{Equivalence ratio, } \phi = \frac{(A/F)_{stoichiometric}}{(A/F)_{actual}} = \frac{17.26}{17.51} = 0.986.$$

Worked Example 4.9 – Working with fuel mixtures
A gaseous fuel mixture comprises 65% methane (CH_4), 25% ethane (C_2H_6) and 10% propane (C_3H_8) by volume.

Determine the air–fuel ratio if the mixture is burnt with stoichiometric air.

Solution

Per mole of fuel:

$$(0.65CH_4 + 0.25C_2H_6 + 0.1C_3H_8) + a(O_2 + 3.76N_2) \rightarrow bCO_2 + cH_2O + dN_2$$

Balancing between reactants and products:

Element	Number of atoms in reactants	Number of atoms in products	Balance result
C	$0.65(1) + 0.25(2) + 0.1(3)$	b	$b = 1.45$
H	$0.65(4) + 0.25(6) + 0.1(8)$	$2c$	$c = 4.9/2 = 2.45$
O	$2a$	$2b + c$	$a = [2(1.45) + 2.45]/2 = 2.675$
N	$2(3.76)(a)$	$2d$	$d = 2(3.76)(2.675)/2 = 10.058$

The balanced equation is, therefore:

$$(0.65CH_4 + 0.25C_2H_6 + 0.1C_3H_8) + 2.675(O_2 + 3.76N_2)$$
$$\rightarrow 1.45CO_2 + 2.45H_2O + 10.058N_2$$

The air–fuel ratio is given by:

$$(A/F) = m_{air}/m_{fuel} = \sum(nM_{air}) / \sum(nM_{fuel})$$

$$= \frac{(2.675)(1 + 3.76)(29)}{[(0.65 \times 1) + (0.25 \times 2) + (0.1 \times 3)](12) + [(0.65 \times 4) + (0.25 \times 6) + (0.1 \times 8)](1)}$$

$$= 16.6$$

Worked Example 4.10 – Use of SFEE (non-stoichiometric conditions)

Gaseous octane (C_8H_{18}) is burnt with 250% excess air in a steady-flow process.

If the reactants enter the combustion chamber at 25 °C and leave at 800 K, determine the heat release (kJ/kmol of fuel).

Solution

The stoichiometric equation for the combustion of octane is:

$$C_8H_{18}(g) + 12.5(O_2 + 3.76N_2) \rightarrow 8CO_2 + 9H_2O + 47N_2$$

250% excess air is 350% stoichiometric air, so the actual equation is:

$$C_8H_{18}(g) + 12.5(3.5)(O_2 + 3.76N_2) \rightarrow 8CO_2 + 9H_2O + 12.5(2.5)O_2 + (2.5)47N_2$$

From Table B.1 (see Appendix B) for the *reactants*:

$$\bar{h}^o_{f,octane}(g) = -208\,450\,kJ/kmol$$

$$\bar{h}^o_{f,O_2} = 0$$

$$\bar{h}^o_{f,N_2} = 0$$

From Table B.1 (see Appendix B) for the *products*:

$$\bar{h}^\circ_{f, CO_2} = -393\,520\,kJ/kmol$$

$$\bar{h}^\circ_{f, H_2O} = -241\,820\,kJ/kmol$$

From Table B.2 (see Appendix B) for the *products* at 800 K:

$$\left(\bar{h} - \bar{h}^\circ\right)_{CO_2} = 22\,815\,kJ/kmol$$

$$\left(\bar{h} - \bar{h}^\circ\right)_{H_2O} = 17\,991\,kJ/kmol$$

$$\left(\bar{h} - \bar{h}^\circ\right)_{N_2} = 15\,046\,kJ/kmol$$

$$\left(\bar{h} - \bar{h}^\circ\right)_{O_2} = 15\,841\,kJ/kmol$$

Applying the SFEE:

$$Q - W = \sum_{\text{products}} n_i [\bar{h}^\circ_f + (\bar{h} - \bar{h}^\circ)]_i - \sum_{\text{reactants}} n_i [\bar{h}^\circ_f + (\bar{h} - \bar{h}^\circ)]_i$$

$$= [(8)(-393\,520 + 22\,815) + (9)(-241\,820 + 17\,991)$$
$$+ (31.25)(0 + 15\,841) + (117.5)(0 + 15\,046)]_{\text{products}}$$

$$-[(-208\,450) + (43.75)(0) + (43.75)(3.76)(0)]_{\text{reactants}}$$

$$= -2\,516\,238\,kJ/kmol \text{ of fuel}$$

Worked Example 4.11 – Use of NFEE (stoichiometric conditions)

In a constant-volume process, 1 kmol of butane (C_4H_{10}) is combusted in stoichiometric proportions with *oxygen*.

The reactants were originally at 298 K. The products are cooled to 600 K.

Assuming the work done is zero, determine the heat release (kJ) during the process.

Solution

The fuel is burnt in oxygen and not air, so we have:

$$C_4H_{10} + aO_2 \rightarrow bCO_2 + cH_2O$$

Balancing between reactants and products:

Element	Number of atoms in reactants	Number of atoms in products	Balance result
C	4	b	$b = 4$
H	10	$2c$	$c = 10/2 = 5$
O	$2a$	$2b + c$	$a = [2(4) + 5]/2 = 6.5$

The balanced equation is, therefore:

$$C_4H_{10} + 6.5O_2 \rightarrow 4CO_2 + 5H_2O$$

From Table B.1 for the *reactants*:

$$\bar{h}^o_{f,butane} = -126\,150\,kJ/kmol$$

$$\bar{h}^o_{f,O_2} = 0$$

From Table B.1 for the *products*:

$$\bar{h}^o_{f,CO_2} = -393\,520\,kJ/kmol$$

$$\bar{h}^o_{f,H_2O} = -241\,820\,kJ/kmol$$

From Table B.2 for the *products* at 600 K:

$$(\bar{h} - \bar{h}^o)_{CO_2} = 12\,916\,kJ/kmol$$

$$(\bar{h} - \bar{h}^o)_{H_2O} = 10\,498\,kJ/kmol$$

$$Q - W = \sum_{products} n_i[\bar{h}^o_f + (\bar{h} + \bar{h}^o) - R_oT]_i$$

$$- \sum_{reactants} n_i[\bar{h}^o_f + (\bar{h} + \bar{h}^o) - R_oT]_i$$

$$= [4(-393\,520 + 12\,916) + 5(-241\,820 + 10\,498)$$

$$- (9 \times 8.314 \times 600)]_{products}$$

$$- [1(-126\,150) - (7.5 \times 8.314 \times 298)]_{reactants}$$

$$= -2\,579\,190\,kJ$$

Worked Example 4.12 – Adiabatic flame temperature

Methane (CH_4) is burned with 30% excess air at 25 °C, 1 atm.
Determine the adiabatic flame temperature (K) for the reaction.

Solution

The stoichiometric equation for the combustion of methane is:

$$CH_4 + 2(O_2 + 3.76N_2) \rightarrow CO_2 + 2H_2O + 7.52N_2$$

30% excess air is 130% stoichiometric air, so the actual equation is:

$$CH_4 + (1.3)2(O_2 + 3.76N_2) \rightarrow CO_2 + 2H_2O + 0.6O_2 + (1.3)7.52N_2$$

From Table B.1 for the *reactants*:

$$\bar{h}^o_{f,methane} = -74\,850\,kJ/kmol$$

$$\bar{h}^o_{f,O_2} = 0, \quad h^o_{f,N_2} = 0$$

From Table B.1 for the *products*:

$$\bar{h}^o_{f,\ CO_2} = -393\ 520\ \text{kJ/kmol}$$

$$\bar{h}^o_{f,\ H_2O} = -241\ 820\ \text{kJ/kmol}$$

Calculation of the product enthalpy requires knowledge of their temperature. Guess $T_{products} = 2000$ K.
From Table B.2 for the *products*:

$\bar{h} - \bar{h}^o$(kJ/kmol)			
CO_2	H_2O	N_2	O_2
91 450	72 689	56 141	59 199

Under adiabatic conditions:

$$\sum_{\text{reactants}} n_i \bar{h}_i = \sum_{\text{products}} n_i \bar{h}_i$$

$$[-74\ 850 + 0 + 0]_{\text{reactants}} = [(1)(-393\ 520 + 91\ 450)$$

$$+ (2)(-241\ 820 + 72\ 689) + (0.6)(0 + 59\ 199) + (9.78)(0 + 56\ 141)]_{\text{products}}$$

$$= -55\ 754\ \text{kJ/kmol}$$

That is, the product temperature guess is *too hot!*
Guess $T_{products} = 1800$ K.
From Table B.2 for the *products*:

$\bar{h} - \bar{h}^o$(kJ/kmol)			
CO_2	H_2O	N_2	O_2
79 442	62 609	48 982	51 689

Using

$$\sum_{\text{reactants}} n_i \bar{h}_i = \sum_{\text{products}} n_i \bar{h}_i$$

$$[-74\ 850 + 0 + 0]_{\text{reactants}}$$

$$= [(1)(-393\ 520 + 79\ 442) + (2)(-241\ 820 + 62\ 609)$$

$$+ (0.6)(0 + 51\ 689) + (9.78)(0 + 48\ 982)]_{\text{products}}$$

$$= -162\ 443\ \text{kJ/kmol}$$

That is, the product temperature guess is *too cold!*

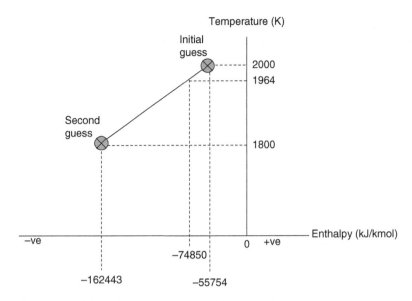

Figure 4.3 Adiabatic flame temperature graphical solution.

Use linear interpolation between 1800 and 2000 K to arrive at a solution.

$$T_{products} = 2000 - \left\{ \frac{[74\ 850 - 55\ 754]}{[162\ 443 - 55\ 754)]} \times (2000 - 1800) \right\} = 1964 \text{ K}$$

Figure 4.3 shows the solution in graphical form.

4.9 Tutorial Problems

Note: When attempting the questions below, a reasonable approach should be made with respect to rounding of values to achieve the final result. This is especially applicable to enthalpy-based problems.

4.1 Determine the molar mass (kg/kmol) of the following molecules:
   (a) Carbon monoxide (CO).
   (b) Sulphur dioxide (SO_2).
   (c) Benzene (C_6H_6).
   (d) Ethyl alcohol (C_2H_5OH).

   [Answers: 28 kg/kmol, 64 kg/kmol, 78 kg/kmol, 46 kg/kmol]

4.2 A biogas resulting from a process known as *anaerobic digestion* (a little like composting without air) comprises 60% methane (CH_4) and 40% carbon dioxide (CO_2) per kmol of fuel.
   Determine the resulting molar mass (kg/kmol) of the gas mixture.

   [Answer: 27.2 kg/kmol]

4.3 Dodecane ($C_{12}H_{26}$) is often used as an approximate model for commercial diesel fuel. If the density of the fuel is 755 kg/m³, determine the number of moles per litre of fuel.

[Answer: 4.44 moles]

4.4 Produce balanced stoichiometric combustion equations for the following fuels with air:
(a) Pentane (C_5H_{12}).
(b) Ethyl alcohol (C_2H_5OH).
(c) Nitromethane ($C_2H_3NO_2$).
Express the air–fuel ratio in each case.

[Answers: 15.34, 9.0, 3.31]

4.5 A kilomole of methane (CH_4) is burnt with dry air that contains 3 kmol of O_2. Calculate the % excess air, the air–fuel ratio and the molar fraction (%) of oxygen in the products.

[Answers: 50%, 25.88, 6.5%]

4.6 A butane (C_4H_{10})–air mixture has an A/F ratio of 21. Determine:
(a) The stoichiometric A/F ratio.
(b) The % excess air and the % stoichiometric air.
(c) Whether the mixture is *rich* or *lean* (justify your answer).
(d) The volume percentage of CO_2 in the products.
(e) The CO_2/fuel mass ratio.

[Answers: 15.47, 35.74%, 135.74%, lean, 8.98%, 3.03 kg_{CO_2}/kg_{fuel}]

4.7 Propane (C_3H_8) is burned with dry air. A volumetric analysis of the products on a dry basis gives 11% CO_2, 4.5% O_2 and 84.5% N_2. Calculate the air–fuel ratio and the % excess air.

[Answers: 19.2, 22%]

4.8 An unknown hydrocarbon fuel combusts with dry air. The resulting products have the following dry volumetric analysis: 7.7% CO_2, 9.8% O_2 and 82.5% N_2. Estimate the fuel composition in terms of its carbon/hydrogen atomic ratio. Calculate the % excess air.

[Answers: 0.43, 81%]

4.9 Carbon is burnt with pure oxygen to form carbon dioxide in a steady-flow reaction. Assuming isothermal and isobaric standard state conditions, calculate the energy evolved.

[Answer: −393 520 kJ/kmol of fuel]

4.10 Assuming steady flow, isothermal conditions (25 °C, 1 atm), verify that the enthalpy of combustion for a stoichiometric mixture of gaseous propane (C_3H_8) and air is −2 220 000 kJ/kmol.

What is the effect on the enthalpy of combustion (kJ/kmol) if the propane enters the combustion process in a *liquid* state?

Take the enthalpy of vaporization of propane to be 15 060 kJ/kmol.

Assume product water to be in a liquid state in both cases.

[Answer: −15060 kJ/kmol of fuel]

4.11 Gaseous methyl alcohol or *methanol* (CH_3OH) and air in stoichiometric proportions at 25 °C, 1 atm are burnt in a steady-flow combustion chamber.

If the products leave the chamber at 600 K and 1 atm, calculate the heat release.

[Answer: Approximately −542 000 kJ/kmol of fuel]

4.12 *Liquid* benzene (C_6H_6) at 25 °C, 1 atm is burnt with 95% stoichiometric air in a steady-flow combustor. The products contain the stoichiometric amount of water. The products also contain both carbon dioxide (CO_2) *and* carbon monoxide (CO). If the enthalpy of vaporization of benzene is 33 830 kJ/kmol, determine:

(a) The carbon/hydrogen mass ratio for the fuel.
(b) The molar fraction of CO in the products (%).
(c) The heat released (kJ/kmol of fuel) if the products leave the combustion chamber at 1000 K.

[Answers: 12, 2.1%, approximately −2 079 000 kJ/kmol of fuel]

4.13 Butane (C_4H_{10}) at 25 °C is burnt in a steady-flow combustion chamber with 50% excess air. The air supply is preheated to 400 K. However, only 97% of the carbon is converted to CO_2 with the remaining 3% of carbon appearing as carbon monoxide (CO) in the products. If the products leave the chamber at 1000 K, calculate the heat transfer.

Assume the pressure remains constant at 1 atm.

[Answer: approximately −1 638 000 kJ/kmol of fuel]

4.14 Nitromethane (CH_3NO_2) is a *liquid* fuel with a structure based on the simple methane molecule.

Assuming stoichiometric proportions with air, a steady-flow analysis and isothermal conditions, using the data below, determine for the methane derivative:

(a) The standard enthalpy of combustion (kJ/kmol of fuel), assuming the product water is in a liquid state.
(b) The product/reactant molar ratio for the reaction.

Nitromethane (liquid) data:

Enthalpy of formation ($h°_f$) at 25 °C, 1 atm: −110 000 kJ/kmol.

[Answers: −712 265 kJ/kmol, 1.27 K]

4.15 Ethane (C_2H_6) at 25 °C undergoes complete combustion with air at 400 K, 1 atm in a steady-flow, insulated combustor.

Determine the adiabatic flame temperature for 50% excess air.

[Answer: 1894 K]

4.16 An attempt to reduce nitric oxide (NO) emissions in a flue gas uses ammonia in the following reaction:

$$6NO + 4NH_3 \leftrightarrows 5N_2 + 6H_2O$$

Determine the mass (kg) of ammonia required per tonne of NO.

[Answer: 378 kg]

4.17 A large coal-fired power plant emits 30 kt of sulphur dioxide (SO_2) per annum. It is proposed to capture the emission using calcium carbonate according to the following basic reaction:

$$2CaCO_3 + 2SO_2 + O_2 \leftrightarrows 2CaSO_4 + 2CO_2$$

Determine how much carbon dioxide (CO_2) is generated per annum as a consequence of the SO_2 clean-up.

[Answer: 20 625 tonnes per annum]

5

Control of Particulates

5.1 Overview

The term *particulate matter* (PM) or *particulates* describes solid or liquid particles released into the atmosphere from either man-made or natural sources.

At low levels in the atmosphere, particulates are harmful to human health whilst at higher atmospheric levels they may alter the earth's heat balance due to selective absorption and reflection of electromagnetic radiation.

Natural particulate sources include volcanoes and sea spray.

Significant man-made sources include solid fuel, for example, coal or biomass-fired electricity generation plants and diesel engines.

This chapter will introduce some of the significant parameters associated with particle dynamics and particle capture and will go on to apply these to common particulate collection systems such as cyclones and electrostatic precipitators.

Learning Outcomes

- To understand the nature of particulates.
- To be familiar with basic particle collection efficiency and associated performance definitions.
- To be familiar with basic particle forces and associated dimensionless parameters.
- To be familiar with particulate collection devices such as gravity settlers, cyclones, ESPs etc., and to use the mass conservation law to model performance.

5.2 Some Particle Dynamics

5.2.1 Nature of Particulates

The most commonly used unit for expressing particulate concentration in the environment is weight (*of particulate*) per unit volume (*of environment*), measured in micrograms per cubic metre ($\mu g/m^3$); however, particulate loading concentrations are sometimes expressed in parts per million by weight (ppm_w).

Conventional and Alternative Power Generation: Thermodynamics, Mitigation and Sustainability, First Edition. Neil Packer and Tarik Al-Shemmeri.

Particulate matter is characterized by size. Although not necessarily spherical, it is commonly specified in terms of an equivalent mean diameter (d_p, m). Due to the small scale, the diameter is often quoted using the sub-multiple micrometre (μm) or 'micron'.

A typical size range for particulate matter is 0.01–100 μm.

The size of raindrops, for scale comparison, is typically of the order of 1000 μm and bacteria are in the range 0.4–400 μm.

Particulates with a diameter of < 1 μm generally do not perceive the surrounding fluid as a flow continuum but are affected by collisions with *individual* fluid molecules.

Usually, particulate matter of a given size is abbreviated to PM_d, where d is a diameter in microns. For example, PM_{10} relates to particulates with a diameter of 10 microns or less.

Particulates with a diameter of between 1 and 10 μm form stable suspensions and are commonly termed *smoke*. The term *dust* is often applied to the smallest particles of diameter 0.01 to 0.1 μm.

The human upper air tract is reasonably efficient in filtering out particulates with a diameter *in excess* of 10 μm (PM_{10}).

If particulates with a diameter of *less than* 10 μm enter the lungs, they have a tendency to remain resident in the alveoli or air sacs, with the smallest able to enter the bloodstream.

Particulate-polluted gas streams are rarely homogenous, containing a range of particle sizes. Particulate size distributions are normally characterized by either arithmetic or geometric median or mean diameter.

The results of particulate measurements are commonly illustrated by plotting frequency distribution or cumulative distribution curves.

5.2.2 Stokes's Law and Terminal Velocity

When a particulate moves through a gas, the fluid viscosity produces a force, known as *drag*, on the particulate in the opposite direction to the relative velocity ($\bar{V}_r$).

The *drag force* (F_D) on a spherical particle is given by:

$$F_D = C_D \left(\frac{\rho_g}{2} \right) \left(\frac{\pi d_p^2}{4} \right) \bar{V}_r^2 \tag{5.1}$$

where C_D is the drag coefficient.

Consider a spherical particle settling through a fluid *at rest* under the influence of gravity (see Figure 5.1). The particle will accelerate until the frictional drag force +

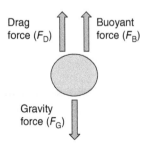

Figure 5.1 Particle forces on a body falling under the effects of gravity.

buoyancy force equals the gravity force. At this point, the particle continues to fall at a constant velocity known as the *terminal settling velocity* ($\bar{V}_{ts}$).

Performing a force balance on the particle, the resultant force is given by:

$$F_R = F_G - F_B - F_D \tag{5.2}$$

where:

$$F_R = \rho_p V_p \left(\frac{d\bar{V}_p}{dt} \right) \tag{5.3}$$

$$F_G = \rho_p V_p g \tag{5.4}$$

$$F_B = \rho_g V_p g \tag{5.5}$$

$$F_D = C_D \left(\frac{\rho_g}{2} \right) \left(\frac{\pi d_p^2}{4} \right) \bar{V}_r^2 \tag{5.6}$$

After substituting into Equation (5.2):

$$\rho_p V_p \left(\frac{d\bar{V}_p}{dt} \right) = \rho_p V_p g - \rho_g V_p g - C_D \left(\frac{\rho_g}{2} \right) \left(\frac{\pi d_p^2}{4} \right) \bar{V}_r^2 \tag{5.7}$$

Dividing through by $\rho_p V_p$ and noting that, for a spherical particle:

$$\frac{A_p}{\rho_p V_p} = \frac{\left(\frac{\pi d_p^2}{4} \right)}{(\pi d_p^3/6)\rho_p} = \frac{3}{2\rho_p d_p} \tag{5.8}$$

Assuming still air i.e. $\bar{V}_P = \bar{V}_r$ and rearranging:

$$\frac{d\bar{V}_p}{dt} = \left[\frac{(\rho_p - \rho_g)}{\rho_p} \right] g - \left[\frac{3 C_D \rho_g \bar{V}_p^2}{4 \rho_p d_p} \right] \tag{5.9}$$

At terminal settling velocity:

$$\frac{d\bar{V}_p}{dt} = 0 \quad \text{i.e., acceleration is zero.}$$

Solving for ($\bar{V}_p$), i.e. the terminal settling velocity ($\bar{V}_{ts}$):

$$\bar{V}_{ts} = \sqrt{\left[\frac{4g(\rho_p - \rho_g)d_p}{3 C_D \rho_g} \right]} \tag{5.10}$$

The drag coefficient, C_D, is a dimensionless number dependent on the shape or form of the particle and the friction between the fluid and the surface (or skin) of the particle.

For *spherical* particles, the approximate graphical relationship with Reynolds number is shown in Figure 5.2.

Values for C_D are found empirically and are usually correlated with Reynolds number (Re) in the following form:

$$C_D = b/Re^n \tag{5.11}$$

Correlations for a range of Re values are given in Table 5.1.

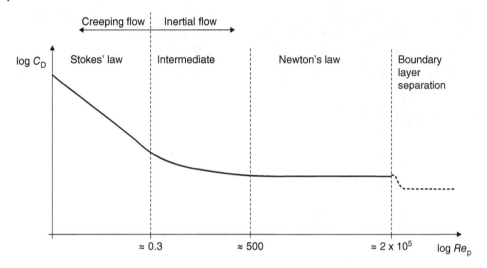

Figure 5.2 Generalized drag coefficient–Reynolds number relationships for spherical bodies. Reproduced with permission from Rhodes, M. (2008) *Introduction to Particle Technology*, 2nd edition, John Wiley & Sons.

Table 5.1 C_D–Re empirical parameters.

Region	Stokes's Law	Intermediate	Newton's Law
Re range	<0.3	0.3 < Re < 500	500 < Re < 2×10^5
C_D	24/Re	$(24/\text{Re})(1 + 0.15\text{Re}^{0.687})$	~0.44

Source: Rhodes, M. (2008) *Introduction to Particle Technology*, 2nd edition, John Wiley & Sons.

Assuming laminar (or Stokes flow) conditions:

$$C_D = 24/\text{Re} = \frac{24\mu_g}{\rho_g \bar{V}_{ts} d_p} \tag{5.12}$$

Substituting into Equation (5.10) gives:

$$\bar{V}_{ts} = \left[\frac{(\rho_p - \rho_g)g d_p^2}{18\mu_g} \right] \tag{5.13}$$

which is known as Stokes's Law.

In the study of air pollution, the particulate density (ρ_p) is usually much greater than that of the fluid (ρ_g) and so the effect of the latter is often ignored.

The Stokes number (Stk), a dimensionless parameter commonly used in particle dynamics, is defined as:

$$\text{Stk} = \left[\frac{d_p^2 \rho_p}{18\mu_g} \right] \times \left[\frac{\bar{V}}{d} \right]_{sys} = \tau \left[\frac{\bar{V}}{d} \right]_{sys} \tag{5.14}$$

where the ratio ($\bar{V}/d$) is a velocity/length characteristic of the system and τ may be regarded as the time constant with respect to the attainment of terminal velocity.

Analyses using Stokes's Law tend to become unreliable for particle sizes below 0.1 µm and above 50 µm.

5.3 Principles of Collection

5.3.1 Collection Surfaces

The efficiency (η_c) of the removal process for a collecting surface is defined as the mass flow rate with which the particles are removed by contact with the collecting surface divided by the mass flow rate of particles approaching the surface in a stream tube of cross-sectional area equal to the collecting surface.

The geometry of a collecting surface comprising *cylindrical fibres* of length L_{fb} and diameter d_{fb} is shown in Figure 5.3.

Using the definition above:

$$\eta_c = \frac{\text{Mass removal rate}}{\bar{V}_r \, c \, d_{fb} L_{fb}} \tag{5.15}$$

where c is the concentration of particles upstream of the collector.

For a collecting surface comprising *spherical drops*, for example, a particulate washer or scrubber, the geometry is shown in Figure 5.4.

Using the definition above:

$$\eta_c = \frac{\text{Mass removal rate}}{\bar{V}_r \, c \left(\frac{\pi d_{sph}^2}{4} \right)} \tag{5.16}$$

In both of the cases, $\bar{V}_r$ is the particle velocity *relative to the collecting surface*.

5.3.2 Collection Devices

Consider the pollution control device shown in Figure 5.5.

The mass flow rate of pollutant upstream of the device is $\dot{V}_o c_o$ and that leaving is $\dot{V}_1 c_1$. Using the mass-based approach to collection efficiency, the following terms can be defined:

$$\text{Collection efficiency, } \eta_c = \frac{\dot{V}_o c_o - \dot{V}_1 c_1}{\dot{V}_o c_o} = 1 - \frac{\dot{V}_1 c_1}{\dot{V}_o c_o} \tag{5.17}$$

Figure 5.3 Cylindrical collector geometry.

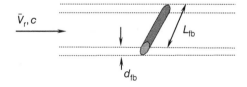

Figure 5.4 Spherical collector geometry.

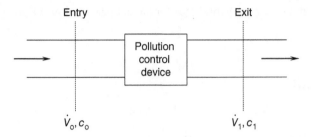

Figure 5.5 Collector operating parameters.

For equal inlet and outlet volume flow rates, this can be simplified to:

$$\eta_c = 1 - (c_1/c_o) \tag{5.18}$$

$$\text{Penetration, } p = 1 - \eta_c = \frac{\dot{V}_1 c_1}{\dot{V}_o c_o} \tag{5.19}$$

$$\text{Decontamination factor, } DF = \frac{1}{\text{penetration}} = \frac{\dot{V}_o c_o}{\dot{V}_1 c_1} \tag{5.20}$$

Particulate control devices are often used in combination.

For N such devices of collection efficiencies $(\eta_1,...,\eta_N)$ *in series* (see Figure 5.6), it can be shown that:

$$\eta_{\text{overall}} = 1 - [(1 - \eta_1)(1 - \eta_2)\,...\,...\,..(1 - \eta_N)] \tag{5.21}$$

For N such collection devices *in parallel* (see Figure 5.7), it can be shown that:

$$\eta_{\text{overall}} = 1 - \left[(1 - \eta_1) \left(\frac{\dot{V}_1}{\dot{V}_{\text{tot}}} \right) + (1 - \eta_2) \left(\frac{\dot{V}_2}{\dot{V}_{\text{tot}}} \right) + ... (1 - \eta_N) \left(\frac{\dot{V}_N}{\dot{V}_{\text{tot}}} \right) \right] \tag{5.22}$$

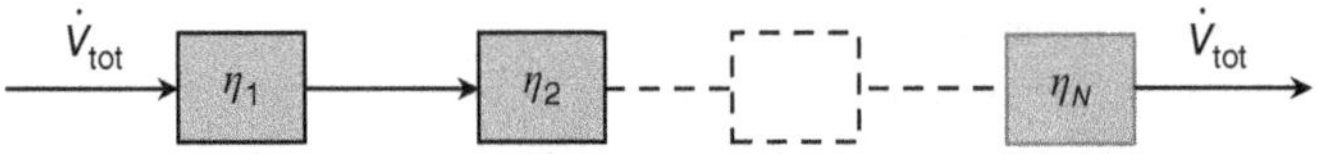

Figure 5.6 Schematic of collectors in series.

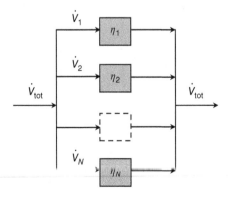

Figure 5.7 Schematic of collectors in parallel.

5.3.3 Fractional Collection Efficiency

As mentioned earlier, particulate pollutant flow streams often comprise a range of particulate sizes. The *fractional efficiency* ($\eta_{c,d}$) is defined as the fraction of particles *of a given size* (*d*) collected compared to the fraction of the same size entering a pollution control device.

The *total collection efficiency* (η_{tot}) of a collection device is given by:

$$\eta_{tot} = \sum (\eta_{c,d} \times mf) \tag{5.23}$$

where mf = the particulate mass fraction of a specific sized particle.

Another term in common usage is the *cut diameter*, defined as the diameter of particle for which the total collection efficiency curve has a value of 0.5, i.e. 50%.

5.4 Control Technologies

Particulate control technologies rely on a range of physical phenomena, such as gravity, centrifugal, electrostatic and impaction forces, to capture and remove particulates from a flow stream. Generally, the smaller the particulate size, the greater the magnitude of force required for its removal.

5.4.1 Gravity Settlers

At its simplest, a gravity settler (or sedimentation chamber) is an expansion in the cross-sectional area of a duct carrying a particulate-laden fluid stream. The expansion results in a reduction in flow velocity. The change in cross-section must be sufficient to allow the particles to attain their terminal velocity.

For the gravity settler shown in Figure 5.8:

$$A = WH \tag{5.24}$$

$$\bar{V}_{ave} = \dot{V}/WH \tag{5.25}$$

Two gas-particle interaction models are in common usage to predict particle collection efficiency. The following assumptions apply to both:

- $\bar{V}_{x,g} = \bar{V}_{x,p} = \bar{V}_{ave}$
- $\bar{V}_{y,p} = \bar{V}_{ts}$

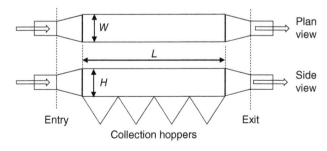

Figure 5.8 Gravity settler overall dimensions.

- Stokes flow
- If a particle falls to the floor, it cannot be re-entrained.

5.4.1.1 Model 1: Unmixed Flow Model

This model assumes there is no mixing mechanism to redistribute *yet-to-be* collected particles across the volume.

The parameters for Model 1 are shown in Figure 5.9.

In time t, at a distance L downstream of the inlet, all the particles of the same diameter have fallen the same distance and the uppermost particles have fallen a distance $(H - y)$.

$$\text{Let } \bar{V}_{ave} = L/t \tag{5.26}$$

and

$$\bar{V}_{ts} = \frac{H - y}{t} \tag{5.27}$$

Rearranging Equation (5.27) and substituting from Equation (5.26):

$$H - y = t\bar{V}_{ts} = (L/\bar{V}_{ave})\bar{V}_{ts} \tag{5.28}$$

Let the average concentration at point L be c_L:

$$c_L = (y/H)c_o \quad \text{i.e. } y = (c_L/c_o)H \tag{5.29}$$

Rearranging and substituting from Equation (5.28):

$$c_L/c_o = 1 - (L/H)(\bar{V}_{ts}/\bar{V}_{ave}) \tag{5.30}$$

Hence, from Equation (5.18):

$$\eta_c = (L/H)(\bar{V}_{ts}/\bar{V}_{ave}) \tag{5.31}$$

Substituting for $\bar{V}_{ts}$ from Equation (5.13):

$$\eta_c = \frac{Lgd_p^2\rho_p}{H\bar{V}_{ave}18\mu_g} \tag{5.32}$$

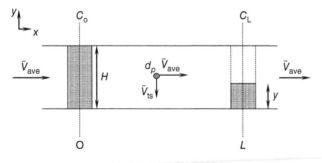

Figure 5.9 Gravity settler 'unmixed' or 'laminar' model parameters.

5.4.1.2 Model 2: Well-mixed Flow Model

This model assumes that, as particles of a certain size fall to the bottom, a mixing mechanism redistributes the remaining particles throughout any given volume.

The parameters for this model are illustrated in Figure 5.10.

Consider a length of settler dx with a uniform concentration c throughout its volume $A.dx$.

Performing a mass balance:

Mass flow rate of particles entering the volume = Mass flow rate of particles leaving the volume + Rate of particle collection in the volume

i.e.
$$c A \bar{V}_{ave} = [c + dc] A \bar{V}_{ave} + c \bar{V}_{ts} W.dx \qquad (5.33)$$

where $W. dx$ is the collection area at the bottom of the settler.

Remembering that: $A = WH$, divide through by $\bar{V}_{ave}$ and A.

Rearranging and separating variables:

$$c - c \frac{\bar{V}_t}{\bar{V}_{ave} H}.dx = c + dc$$

$$c \left(1 - \frac{\bar{V}_{ts}}{\bar{V}_{ave} H}.dx \right) = c + dc$$

$$-\frac{\bar{V}_{ts}}{\bar{V}_{ave} H}.dx = \frac{c + dc}{c} - 1$$

$$-\frac{V_{ts}}{\bar{V}_{ave} H}.dx = \frac{c}{c} + \frac{dc}{c} - 1 = \frac{dc}{c}$$

Integrating between $c = c_o$ and c_L, and $x = 0$ and L:

$$\int_{C_o}^{C_L} \left(\frac{1}{c} \right).dc = \int_0^L - (\bar{V}_{ts}/\bar{V}_{ave} H).dx$$

[Remembering that: $\int \frac{1}{x} dx = \ln x$]

$$\ln c_L - \ln c_o = -(\bar{V}_{ts}/\bar{V}_{ave})(L/H)$$

[Remembering that: $\ln x - \ln y = \ln(x/y)$]

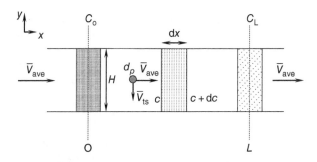

Figure 5.10 Gravity settler 'well-mixed' model parameters.

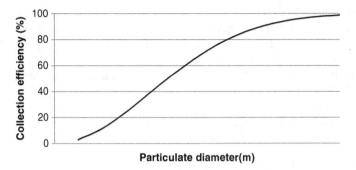

Figure 5.11 Generalized collection efficiency–particle size relationship for well-mixed model assumptions.

Rearranging:

$$c_L/c_o = \exp[-(\bar{V}_{ts}/\bar{V}_{ave})(L/H)]$$

Using the definition of Equation (5.18):

$$\eta_c = 1 - \exp[-(\bar{V}_{ts}/\bar{V}_{ave})(L/H)]$$

Substituting from Equation (5.13):

$$\eta_c = 1 - e^{-\left(\frac{Lgd_p^2 \rho_p}{H\bar{V}_{ave}18\mu_{gas}}\right)} \tag{5.34}$$

Note that:

$$\eta_{\text{well mixed}} = 1 - e^{(-\eta_{\text{laminar flow}})} \tag{5.35}$$

The generalized form of the collection efficiency–particle diameter relationship is shown in Figure 5.11.

5.4.2 Centrifugal Separators or Cyclones

Due to the relatively weak gravitational force, gravity settlers are only suitable for the collection of large particles. In order to be effective for smaller particles, the gravitational force must be replaced by a stronger force.

In the cyclone, the collection force is centrifugal in nature.

Consider a particle in a fluid stream flowing at a radius r in a curved duct of height H (see Figure 5.12).

Assume:

- Solid body rotation.
- The tangential velocity of the gas and particle are equal, i.e. $\bar{V}_g = \bar{V}_p$.
- The gas has no radial velocity component.
- In the radial direction, the centrifugal force is balanced by the viscous drag force.
- Well-mixed conditions, i.e. redistribution of any remaining particles resulting in concentration varying with angle or curvature not radius.

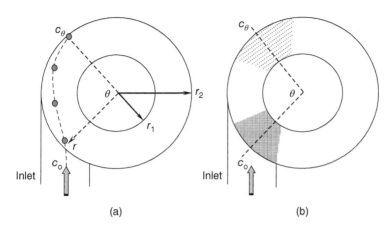

Figure 5.12 Curved duct collection principles. (a) Particle collection path and model parameters; (b) well-mixed model, i.e. redistribution of uncaptured particles across duct.

Let the fluid velocity be a function of radius, i.e. $\bar{V}_g = C_1 r$.
From continuity considerations, the volume flow rate is given by:

$$\dot{V} = \int_{r1}^{r2} \bar{V}_g H.\, dr = C_1 H(r_2^2 - r_1^2)/2 \tag{5.36}$$

Hence

$$C_1 = 2\dot{V}/H(r_2^2 - r_1^2) \tag{5.37}$$

And

$$\bar{V}_g = [2\dot{V}/H(r_2^2 - r_1^2)]r \tag{5.38}$$

Define the average tangential velocity as $\bar{V}_{g,\text{ave}} = \dot{V}/H(r_2 - r_1)$, and an average Stokes's number is thus:

$$\text{Stk}_{\text{ave}} = \frac{\tau \bar{V}_{g,\text{ave}}}{r_2} = \frac{\tau \dot{V}}{r_2 H(r_2 - r_1)} \tag{5.39}$$

Consider the forces acting on a particle at the outer wall (r_2), as shown in Figure 5.13:

➤ A centrifugal force (F_C) acting in the direction of the outer wall.
➤ A retarding viscous drag force (F_D) opposing the motion.

The resultant force is given by:

$$F_R = F_C - F_D \tag{5.40}$$

where $F_R = \rho_p V_p \left(\dfrac{d\bar{V}_p}{dt}\right)$

Figure 5.13 Particle forces in a cyclone.

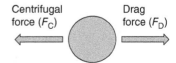

Centrifugal force (F_C) Drag force (F_D)

$$F_C = \rho_p \left(\frac{\pi d_p^3}{6} \right) \left(\frac{\bar{V}_g^2}{r_2} \right)$$

$$F_D = C_D A_p \rho_g \bar{V}_p^2 / 2$$

After substitution:

$$\rho_p V_p d\bar{V}_p / dt = \rho_p \left(\frac{\pi d_p^3}{6} \right) \left(\frac{\bar{V}_g^2}{r_2} \right) - C_D A_p \rho_g \bar{V}_p^2 / 2$$

Make the following assumptions:

Stokes flow, i.e. $C_D = 24/\mathrm{Re} = 24\mu_g / \rho_g \bar{V}_{ts} d_p$

$$A_p = \pi d_p^2 / 4$$

After substitution and rearrangement:

$$\bar{V}_{ts} = \left(\frac{\rho_p d_p^2}{18\mu_g} \right) \left(\frac{\bar{V}_g^2}{r_2} \right) = \frac{\tau \bar{V}_g^2}{r_2} \tag{5.41}$$

A mass balance over an angle θ yields:

Mass flow rate of particles entering volume = Mass flow rate of particles leaving volume + Rate of particle collection in volume

$$c\dot{V} = (c + dc)\dot{V} + c\bar{V}_{ts} A d\theta$$

Substituting from Equation (5.41):

$$c\dot{V} = (c + dc)\dot{V} + c(\tau \bar{V}_g^2 / r_2) r_2 H . d\theta$$

Substituting from Equation (5.38) gives:

$$c\dot{V} = (c + dc)\dot{V} + c[\tau 4\dot{V}^2 r_2^2 / H^2 (r_2^2 - r_1^2)(r_2^2 - r_1^2) r_2] \times (r_2 H) . d\theta$$

Dividing through by $\dot{V}$ and rearranging:

$$c - c \left[\frac{\tau 4\dot{V} r_2^3}{H(r_2^2 - r_1^2)(r_2^2 - r_1^2)} \right] \times (r_2) . d\theta = c + dc$$

$$c(1 - [\tau 4\dot{V} r_2^3 / H(r_2^2 - r_1^2)(r_2^2 - r_1^2)] \times (r_2) . d\theta) = c + dc$$

$$(-[\tau 4\dot{V} r_2^3 / H(r_2^2 - r_1^2)(r_2^2 - r_1^2)] \times (r_2) . d\theta) = \frac{c + dc}{c} - 1$$

$$\left(- \left[\frac{\tau 4\dot{V} r_2^3}{H(r_2^2 - r_1^2)(r_2^2 - r_1^2)} \right] \times (r_2) . d\theta \right) = \frac{c}{c} + \frac{dc}{c} - 1 = \frac{dc}{c}$$

Let $C_2 = \dfrac{4 r_2^3}{(r_2 + r_1)(r_2^2 - r_1^2)}$ \hfill (5.42)

Giving:

$$\frac{dc}{c} = -(\mathrm{Stk}_{ave} \times C_2 . d\theta)$$

Integrating between 0 and θ:

$$\int_{C_0}^{C_\theta} \left(\frac{1}{c}\right).dc = \int_0^\theta (-[\text{Stk}_{\text{ave}} \times C_2.d\theta)$$

[Remembering that: $\int \frac{1}{x}dx = \ln x$]

$$\ln c_\theta - \ln c_0 = -\text{Stk}_{\text{ave}} \times C_2 \times \theta$$

[Remembering that: $\ln x - \ln y = \ln(x/y)$]
Then, rearranging:

$$c_\theta/c_0 = \exp[-\text{Stk}_{\text{ave}} \times C_2 \times \theta]$$

[Remembering that, in general: $\eta_c = 1 - (c_{\text{exit}}/c_{\text{entry}})$]

$$\eta_c = 1 - \exp[-\text{Stk}_{\text{ave}} \times C_2 \times \theta]$$

$$= 1 - \exp[-\text{Stk}_{\text{ave}} \times C_2 \times 2\pi N] \tag{5.43}$$

where N is the number of (*descending*) turns made by the fluid stream.

The equation derived above indicates the relationship between collection efficiency, particle size and cyclone dimensions.

The construction and relative dimensions of a typical top entry/exit cyclone are shown in Figure 5.14.

Shortly after the inlet, the particulate-contaminated gas is forced to turn in the main body of the cyclone.

Due to their inertia, the particles, however, collide with the walls and fall to the bottom of the cone section for collection.

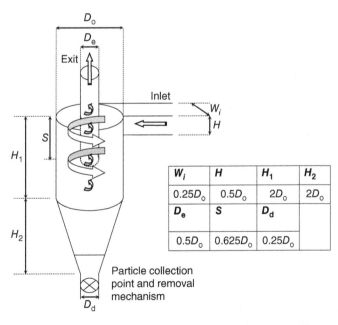

W_i	H	H_1	H_2
$0.25D_o$	$0.5D_o$	$2D_o$	$2D_o$
D_e	S	D_d	
$0.5D_o$	$0.625D_o$	$0.25D_o$	

Particle collection point and removal mechanism

Figure 5.14 Typical or 'classical' cyclone dimensions based on the cylinder diameter (D_o).

As shown in the figure, the gas flow follows a double helical pattern, the outer helix descending, the inner helix rising to the outlet. Only the outer helix contributes to collection.

An empirical expression commonly used for the number of descending turns, N, is:

$$N = (1/H)(2H_1 + H_2) \tag{5.44}$$

In order to maintain high collection efficiency across a range of particle sizes, $\bar{V}_g$ could be increased. This, however, would incur additional power consumption at the system fan.

Alternatively, W_i could be reduced, but to maintain classical cyclone dimensions, this would reduce its overall size and, hence, the flow rate.

One solution to this problem is the multi-cyclone, consisting of many small cyclones in parallel.

5.4.3 Electrostatic Precipitators (ESPs)

Gravitational and centrifugal forces are too weak to provide good collection efficiencies for particles having a diameter of less than 5 μm.

In this case, one alternative would be to use electrostatic forces to drive particles to a collecting surface.

In an ESP, a wire hangs between two plates (see Figure 5.15). One end of the wire is connected through a rectifier to a high voltage supply, setting up an electric field between wire and plate. The other end of the wire is weighted but hangs freely.

Electronic charges leave the wire (the *coronal discharge*) on their journey to the grounded or otherwise charged plates.

Particles carried in the gas flow collect the negative charge as a result of their proximity to the wires and are then driven by electrostatic attraction to the plates (see Figure 5.16).

An electrical field, *EF* (V/m) is said to exist when electrical potential (volts) varies from one point to another in space.

The *permittivity* is a measure of how well a material supports an electric field.

The charge associated with an electron is 1.6×10^{-19} coulombs. The total charge collected by a particle passing through an electric field will depend on the number of electrons collected and will be a multiple of this value.

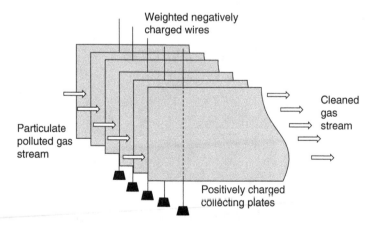

Figure 5.15 Simple plate and wire electrostatic precipitator arrangement.

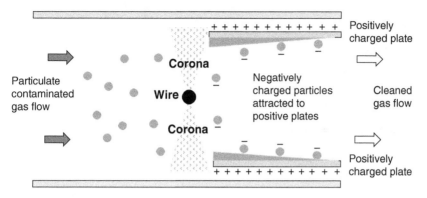

Figure 5.16 Schematic of coronal discharge and particle collection in a single channel.

An often-used relationship for estimating the charge, q (coulombs) accumulated by a particle d_p (m) being charged in an electric field EF_o (V/m) is:

$$q = 3\pi \left(\frac{\varepsilon}{\varepsilon + 2} \right) \varepsilon_o d_p^2 EF_o \qquad (5.45)$$

where ε is a material property known as the *dielectric constant* (or relative permittivity) and ε_o is the permittivity of free space (or vacuum), taken as 8.85×10^{-12} C/Vm.

The electrostatic force (F_E) on a particle is then:

$$F_E = q EF_1 \qquad (5.46)$$

where EF_1 is the local field strength where the particles are collected, i.e. causing the force.

Note that EF_1 may not be equal to EF_o. For example, the particle may acquire its charge in an area of high electric field strength, say close to the wire electrode.

However, for practical purposes, let $EF_o = EF_1 = EF$.

Then, substituting into the above:

$$F_E = 3\pi \left(\frac{\varepsilon}{\varepsilon + 2} \right) \varepsilon_o d_p^2 EF^2 \qquad (5.47)$$

Consider the forces acting on a particle finding itself in an electric field. They are (see Figure 5.17):

➤ An electrostatic attraction force (F_E) in the direction of the voltage potential difference.
➤ A retarding viscous drag force (F_D) opposing the attraction.

The resultant force, F_R, is given by:

$$F_R = F_E - F_D \qquad (5.48)$$

where $F_R = \rho_p V_p \left(\frac{d\bar{V}_p}{dt} \right)$

$$F_E = q EF$$

Figure 5.17 Particle forces in an ESP.

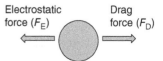

Electrostatic force (F_E)   Drag force (F_D)

$$F_D = C_D A_p \rho_g \bar{V}_p^2 / 2$$

After substitution:

$$\rho_p V_p d\bar{V}_p / dt = qEF - C_D A_p \rho_g \bar{V}_p^2 / 2$$

Making the following assumptions:

Stokes flow, i.e. $C_D = 24/\text{Re} = 24\mu_g / \rho_g \bar{V}_p d_p$

$A_p = \pi d_p^2 / 4$

$qE = 3\pi \left(\dfrac{\varepsilon}{\varepsilon + 2} \right) \varepsilon_0 d_p^2 EF^2$

At zero acceleration, i.e. $\dfrac{d\bar{V}_p}{dt} = 0$

$$0 = 3\pi \left(\frac{\varepsilon}{\varepsilon + 2} \right) \varepsilon_0 d_p^2 EF^2 - \left(\frac{24\mu_g}{\rho_g \bar{V}_p d_p} \right) \left(\frac{\pi d_p^2}{4} \right) (\bar{V}_p^2 / 2)$$

Solving for $\bar{V}_p$, the particle terminal settling velocity, $\bar{V}_{ts}$, is given by:

$$\bar{V}_{ts} = \left[d_p \varepsilon_0 EF^2 \left(\frac{\varepsilon}{\varepsilon + 2} \right) \right] / \mu_g \tag{5.49}$$

In ESP technology, the terminal settling velocity is more commonly termed the particle *drift* or *migration velocity*.

Inspection of the above reveals several important points:

1. $\bar{V}_{ts} \propto EF^2$, so raising the field strength by increasing the voltage or decreasing the wire-to-plate distance will significantly increase drift velocities.

   Unfortunately, this will also increase the incidence of sparking to the point where particle build-up on the plate is disrupted.
2. $\bar{V}_{ts} \propto d_p$, so compared to a cyclone, larger drift velocities would be anticipated at any given particle size.
3. $\bar{V}_{ts}$ is independent of $\bar{V}_{ave}$, hence the residence time of the particle and its probability of capture can be increased by providing a large area.

In terms of collection efficiency analysis, an ESP is similar to that for the well-mixed gravity settler model.

Let an ESP have a plate of width (i.e. height) W, length (in the flow direction) L and a wire-to-plate distance of d. Consider a section dx with a uniform concentration c throughout its volume Adx (see Figure 5.18).

Performing a mass balance:

Mass flow rate of particles entering the volume = Mass flow rate of particles leaving the volume + Rate of particle collection in the volume

That is

$$cA\bar{V}_{ave} = [c + dc]A\bar{V}_{ave} + c\bar{V}_{ts}W.dx$$

where $W.dx$ is the plate collection area.

[Remembering that flow area $A = WD$, divide through by $\bar{V}_{ave}$ and A.]

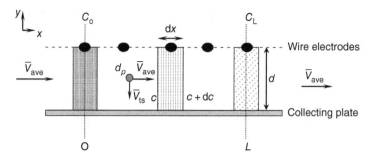

Figure 5.18 Wire and single-plate ESP parameters.

Rearranging and separating variables:

$$c - c\frac{\bar{V}_{ts}}{\bar{V}_{ave}d}.dx = c + dc$$

$$c\left(1 - \frac{\bar{V}_{ts}}{\bar{V}_{ave}d}.dx\right) = c + dc$$

$$-\frac{\bar{V}_{ts}}{\bar{V}_{ave}d}.dx = \frac{c + dc}{c} - 1$$

$$-\frac{V_{ts}}{\bar{V}_{ave}d}.dx = \frac{c}{c} + \frac{dc}{c} - 1 = \frac{dc}{c}$$

Integrating between $c = c_o$ and c_L, and $x = 0$ and L:

$$\int_{C_o}^{C_L}\left(\frac{1}{c}\right).dc = \int_0^L -(\bar{V}_{ts}/\bar{V}_{ave}d).dx$$

[Remembering that: $\int \frac{1}{x}dx = \ln x$]

$$\ln c_L - \ln c_o = -(\bar{V}_{ts}/\bar{V}_{ave})(L/d)$$

[Remembering that: $\ln x - \ln y = \ln(x/y)$]
Then, rearranging:

$$c_L/c_o = \exp[-(\bar{V}_{ts}/\bar{V}_{ave})(L/d)]$$

[Remembering that, in general: $\eta_c = 1 - (c_{exit}/c_{entry})$]

$$\eta_c = 1 - \exp[-(\bar{V}_{ts}/\bar{V}_{ave})(L/d)]$$

At this point, the earlier-derived expression for terminal velocity could be substituted, but instead most ESP analyses take the following approach.

Substituting $\bar{V}_{ave} = \dot{V}/Wd$ and remembering that $A_{plate} = WL$:

$$\eta_c = 1 - \exp\left(-\frac{\bar{V}_{ts}A_{plate}}{\dot{V}}\right) \tag{5.50}$$

However, practical ESPs have collection plates on both sides of the wire electrodes.

Therefore, the total plate collecting area, $A_{\text{total plate}} = 2 \times A_{\text{plate}}$ and the total volume flow rate, $\dot{V}_{\text{total}} = 2\,\dot{V}$.

Hence:

$$\eta_c = 1 - \exp\left(-\frac{\bar{V}_{\text{ts}} A_{\text{total plate}}}{\dot{V}_{\text{total}}}\right) \tag{5.51}$$

ESP performance is less than that predicted by theory due to several factors including re-entrainment of trapped particles during removal via plate rapping and problems with the electrical resistivity of collected particles.

For example, low-resistivity particles will result in small voltage gradients in the collected particles or *cake* formed on the plate and hence, reduced electrostatic forces holding the collected particles to the plate. Again, re-entrainment could result.

Alternatively, high-resistivity particles, resulting from, for example, the combustion of low-sulphur coals, can produce low voltage gradients at the wire, resulting in poor charging of the particles or very high voltage gradients in the cake, leading to thermal energy generation and minor gas explosions, potentially blowing the cake off the plate.

These problems can be overcome by conditioning the gas stream with sulphur trioxide (SO_3) prior to the ESP or separating the charging and collection zones.

Ozone production can also be a problem associated with the field generation, and a specialist catalyst may be needed to reduce concentrations.

5.4.4 Fabric Filters

Filters function by dividing the flow to the point where particles can be captured by *interception*, *impaction* and *diffusion* on fibres and *sieving* by already-collected particles (see Figure 5.19).

(A) *Impaction* – This mechanism is based on the inertia of the particle and its inability to flow around the fibre with the gas flow path (stream line).

(B) *Interception* – The flow path of the particle is such that the fibre is directly in its path.

(C) *Diffusion* – A phenomenon experienced by smaller particles resulting from Brownian motion, i.e. gas molecule bombardment.

(D) *Sieving* – Particles trapped by a filter surface may bridge over the gaps between fibres, with the resulting cake then sieving particles of a size smaller than the bare filter.

Empirically, it can be shown that for *impaction* on a *single cylindrical fibre*, the collection efficiency (η_{fb}) is given by:

$$\eta_{\text{fb}} = [\text{Stk}/(\text{Stk} + 0.85)]^{2.2} \tag{5.52}$$

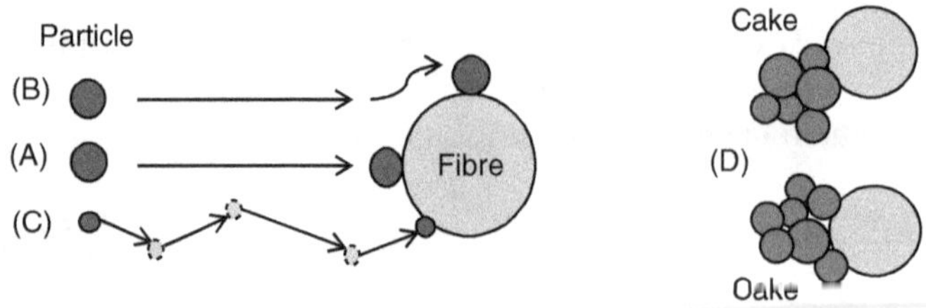

Figure 5.19 Fibre collection mechanisms.

where

$$\text{Stk} = [d_p^2 \rho_p / 18 \mu_g] \times [\bar{V_r} / d_{fb}] \tag{5.53}$$

[Note that the filter *face velocity* is commonly taken as representative of the relative system velocity ($\bar{V_r}$) in the calculation of Stokes's number.]

The total volume occupied by the filter comprises the fibre volume and the void volume (v).

The percentage or fraction of the overall filter volume that is composed of solid fibres is termed the *fibre solid fraction* (f_f)

If the filter solid is regarded as being composed of one long length of fibre (L_{fb}) *per unit volume of filter* then:

$$f_f = \frac{\pi d_{fb}^2 L_{fb}}{4} \tag{5.54}$$

Rearranging:

$$L_{fb} = \frac{4}{\pi} \times \frac{f_f}{d_{fb}^2} \tag{5.55}$$

The percentage or fraction of the overall filter volume that is free space is commonly termed the *voidage* or *porosity*.

$$\text{Porosity} = 1 - f_f \tag{5.56}$$

From continuity considerations:

$$\bar{V}_{face} A_{face} = \bar{V}_v A_v \tag{5.57}$$

where $\bar{V}_v$ is the void velocity and A_v is the void area.

In terms of flow, the porosity may also be regarded as the fraction of total cross-sectional (face) area of the filter that is free space or void, i.e.

$$A_{face} = \frac{A_v}{1 - f_f} \tag{5.58}$$

Consider a particulate-laden air stream passing through a filter of thickness H, as shown in Figure 5.20.

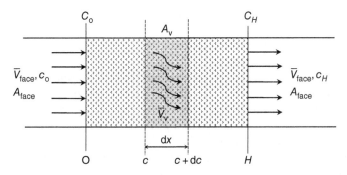

Figure 5.20 Fabric filter parameters.

A mass balance for the particles across section dx gives:

Mass flow rate of particles entering filter = Mass flow rate of particles leaving filter + Mass collection rate in filter

$$c\bar{V}_{face}A_{face} = (c + dc)\bar{V}_{face}A_{face} + (\bar{V}_v d_{fb} L_{fb} c\eta_{fb} A_{face}.dx)$$

Substituting from the continuity equation:

$$c\bar{V}_v A_v = (c + dc)\bar{V}_v A_v + (\bar{V}_v d_{fb} L_{fb} c\eta_{fb} A_{face}.dx)$$

Substituting for A_{face} (Equation (5.58)), L_{fb} (Equation (5.55)) and cancelling:

$$c = (c + dc) + \frac{4}{\pi} \times \frac{f_f}{1 - f_f} \times c \times \eta_{fb} \times \frac{1}{d_{fb}}.dx$$

Rearranging and separating variables:

$$c - c\frac{f_f}{1 - f_f} \times \frac{4}{\pi} \times \frac{\eta_{fb}}{d_{fb}}.dx = c + dc$$

$$c\left(1 - \frac{f_f}{1 - f_f} \times \frac{4}{\pi} \times \frac{\eta_{fb}}{d_{fb}}.dx\right) = c + dc$$

$$-\frac{f_f}{1 - f_f} \times \frac{4}{\pi} \times \frac{\eta_{fb}}{d_{fb}}.dx = \frac{c + dc}{c} - 1$$

$$-\frac{f_f}{1 - f_f} \times \frac{4}{\pi} \times \frac{\eta_{fb}}{d_{fb}}.dx = \frac{c}{c} + \frac{dc}{c} - 1 = \frac{dc}{c}$$

Integrating between $c = c_o$ and c_H, and $x = 0$ and H:

$$\int_{C_o}^{C_H} \left(\frac{1}{c}\right).dc = \int_0^H -\left(\frac{f_f}{1 - f_f} \times \frac{4}{\pi} \times \frac{\eta_{fb}}{d_{fb}}\right).dx$$

[Remembering that: $\int \frac{1}{x} dx = \ln x$]

$$\ln c_H - \ln c_o = -\frac{f_f}{1 - f_f} \times \frac{4}{\pi} \times \frac{\eta_{fb}}{d_{fb}} H$$

[Remembering that: $\ln x - \ln y = \ln(x/y)$]
Then, rearranging:

$$c_H / c_O = \exp\left(-\frac{f_f}{1 - f_f} \times \frac{4}{\pi} \times \frac{\eta_{fb}}{d_{fb}} H\right)$$

[Remembering that, in general: $\eta_c = 1 - (c_{exit}/c_{entry})$]

$$\eta_c = 1 - \exp\left[-\left(\frac{f_f}{1 - f_f}\right) \times \left(\frac{4}{\pi}\right) \times \left(\frac{\eta_{fb}}{d_{fb}}\right) H\right] \quad (5.59)$$

The most large-scale common use of these principles can be seen in the *baghouse filter*. The filter medium in the type of baghouse illustrated in Figure 5.21 takes the form of an array of bags or *inverted socks* open at the bottom to the particulate-contaminated air.

The only way for the air to reach the exit point is to flow up inside the sock and through the filter medium, leaving a layer of particulates on the inside of the sock.

The 'toe' end of each sock is connected to a bar that is periodically agitated, causing the captured particulates to fall out of the sock to be collected in a hopper beneath. Alternatively, some baghouses use compressed air to dislodge collected particulates.

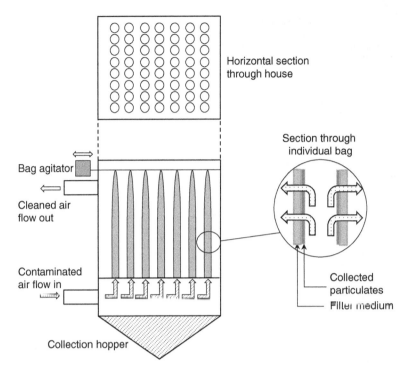

Figure 5.21 Baghouse filter operation.

5.4.5 Spray Chambers and Scrubbers

Spray chambers, washers and scrubbers operate by dividing the flow of a particle-laden gas with multiple small droplets, the fine particles adhering to the passing liquid drop.

This process occurs naturally when a rain shower encounters a dust-laden atmosphere and is sometimes known as *washout* or *scavenging*.

Consider the *counter-flow* spray contact chamber shown in Figure 5.22, with the scrubbing liquid entering at the top and the particulate-laden gas stream entering at the bottom of the chamber.

Assume:

- The diameter of the particles is constant and the concentration varies only with height.
- The flow rates of gas and liquid are constant.
- The particle and gas flow velocities are equivalent, i.e. $\bar{V}_p = \bar{V}_g$
- Stokes flow.
- The scrubbing liquid drops are moving at terminal velocity, i.e. $\bar{V}_{drop} = \bar{V}_{ts}$.
- The system relative velocity $\bar{V}_r = \bar{V}_g - (-\bar{V}_{drop}) = \bar{V}_g + \bar{V}_{drop}$.

Let the volume flow rate of scrubbing liquid ($\dot{V}_L$) be given by:

$$\dot{V}_L = N_{drop}(\pi d^3_{drop}/6)\bar{V}_{drop}A \qquad (5.60)$$

where N_{drop} is the number of collecting drops per unit volume of gas.

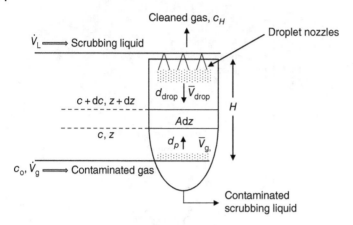

Figure 5.22 Spray chamber parameters.

Considering *impaction only,* it has been shown empirically that the single spherical drop collection efficiency (η_{sph}) is given by:

$$\eta_{sph} = [Stk/(Stk + 0.7)]^2 \tag{5.61}$$

where

$$Stk = [d_p^2 \rho_p / 18\mu_g] \times [\bar{V}_r / d_{drop}] \tag{5.62}$$

Consider the differential volume, $A.dz$, of the spray chamber shown in Figure 5.22. Performing a mass balance:

Mass flow rate of particles entering the volume = Mass flow rate of particles leaving the volume + Rate of particle collection in the volume

That is

$$cA\bar{V}_g = (c + dc)A\bar{V}_g + \eta_{sph}c(\bar{V}_r)N_{drop}(\pi d_{drop}^2/4)A.dz$$

Substituting for N_{drop} (Equation (5.60)) from the above:

$$ScA\bar{V}_g = (c + dc)A\bar{V}_g + \eta_{sph}c\left(\bar{V}_r \times \frac{\dot{V}_L}{\bar{V}_{drop}A} \times \frac{6}{\pi d_{drop}^3}\right) \times (\pi d_{drop}^2/4)A.dz$$

Remembering that: $A\bar{V}_g = \dot{V}_g$ and $\bar{V}_r = \bar{V}_g + \bar{V}_{drop}$

$$c\dot{V}_g = (c + dc)\dot{V}_g + \eta_{sph}c\left((\bar{V}_g + \bar{V}_{drop}) \times \frac{\dot{V}_L}{\bar{V}_{drop}A} \times \frac{6}{\pi d_{drop}^3}\right) \times (\pi d_{drop}^2/4)A.dz$$

Dividing through by $\dot{V}_g$, tidying up all area and volume terms and rearranging:

$$c - \left[c\eta_{sph} \left(\frac{\bar{V}_g + \bar{V}_{drop}}{\bar{V}_{drop}} \right) \left(\frac{\dot{V}_L}{\dot{V}_{gas}} \right) 1.5 \frac{1}{d_{drop}} .dz \right] = c + dc$$

$$c \left[1 - \eta_{sph} \left(\frac{\bar{V}_g + \bar{V}_{drop}}{\bar{V}_{drop}} \right) \left(\frac{\dot{V}_L}{\dot{V}_g} \right) 1.5 \frac{1}{d_{drop}} dz \right] - c \mid dc$$

$$-\eta_{\text{sph}} \left(\frac{\bar{V}_{\text{g}} + \bar{V}_{\text{drop}}}{\bar{V}_{\text{drop}}} \right) \left(\frac{\dot{V}_{\text{L}}}{\dot{V}_{\text{g}}} \right) 1.5 \frac{1}{d_{\text{drop}}}. dz = \frac{c + dc}{c} - 1$$

$$-\eta_{\text{sph}} \left(\frac{\bar{V}_{\text{g}} + \bar{V}_{\text{drop}}}{\bar{V}_{\text{drop}}} \right) \left(\frac{\dot{V}_{\text{L}}}{\dot{V}_{\text{g}}} \right) 1.5 \frac{1}{d_{\text{drop}}}. dz = \frac{c}{c} + \frac{dc}{c} - 1 = \frac{dc}{c}$$

Integrating between $c = c_{\text{o}}$ and c_H, and $x = 0$ and H:

$$\int_{C_{\text{o}}}^{C_H} \left(\frac{1}{c} \right).dc = \int_{\text{o}}^{H} -\eta_{\text{sph}} \left(\frac{\bar{V}_{\text{g}} + \bar{V}_{\text{drop}}}{\bar{V}_{\text{drop}}} \right) \left(\frac{\dot{V}_{\text{L}}}{\dot{V}_{\text{g}}} \right) 1.5 \left(\frac{1}{d_{\text{drop}}} \right).dz$$

[Remembering that: $\int \frac{1}{x} dx = \ln x$]

$$\ln c_H - \ln c_O = -\eta_{\text{sph}} \left(\frac{\bar{V}_{\text{g}} + \bar{V}_{\text{drop}}}{\bar{V}_{\text{drop}}} \right) \left(\frac{\dot{V}_{\text{L}}}{\dot{V}_{\text{g}}} \right) 1.5 \frac{H}{d_{\text{drop}}}$$

[Remembering that: $\ln x - \ln y = \ln(x/y)$]
Then, rearranging:

$$c_H/c_{\text{o}} = \exp \left[-\eta_{\text{sph}} \left(\frac{\bar{V}_{\text{g}} + \bar{V}_{\text{drop}}}{\bar{V}_{\text{drop}}} \right) \left(\frac{\dot{V}_{\text{L}}}{\dot{V}_{\text{g}}} \right) 1.5 \left(\frac{H}{d_{\text{drop}}} \right) \right]$$

[Remembering that, in general: $\eta_c = 1 - (c_{\text{exit}}/c_{\text{entry}})$]

$$\eta_c = 1 - \exp \left[-\eta_{\text{sph}} \left(\frac{\bar{V}_{\text{g}} + \bar{V}_{\text{drop}}}{\bar{V}_{\text{drop}}} \right) \left(\frac{\dot{V}_{\text{L}}}{\dot{V}_{\text{g}}} \right) 1.5 \left(\frac{H}{d_{\text{drop}}} \right) \right] \tag{5.63}$$

5.5 Worked Examples

Worked Example 5.1 – Stokes's Law
A particulate with a diameter of 25 μm falls in air.
Assuming Stokes flow, use the data below to determine its terminal velocity (m/s).

Data
Particle density: 1800 kg/m³, air dynamic viscosity: 1.8×10^{-5} kg/ms, air density: 1.2 kg/m³.

Solution
Given: $\rho_{\text{p}} = 1800$ kg/m³, $\rho_{\text{g}} = 1.2$ kg/m³, $\mu_{\text{g}} = 1.8 \times 10^{-5}$ kg/ms, $d_{\text{p}} = 25 \times 10^{-6}$ m.

Find $\bar{V}_{\text{ts}}$.

$$\bar{V}_{\text{ts}} = \left[\frac{(\rho_{\text{p}} - \rho_{\text{g}})g d_{\text{p}}^2}{18\mu_{\text{g}}} \right]$$

$$= \left[\frac{(1800 - 1.2) \times 9.81 \times (25 \times 10^{-6})^2}{18 \times 1.8 \times 10^{-5}} \right]$$

$$= 0.034 \text{ m/s}$$

Worked Example 5.2 – Collector arrangements and cut diameter

A filter can be modelled by a single *row of parallel cylindrical fibres*, each having a diameter of 400 μm in a simple weave pattern, with the distance between adjacent fibres being 100 μm.

The gas velocity through the cloth is 0.5 m/s. Using the data below, estimate:

(a) The collection efficiency for 15 μm particles.
(b) The cut diameter for the filter.
(c) The efficiency for a filter comprising two rows of parallel cylindrical fibres.

Take the dynamic viscosity of air to be 1.8×10^{-5} kg/ms and the particle density as 1800 kg/m³. Ignore the effect of air density.

Solution

Given: $\rho_p = 1800$ kg/m³, $\mu_g = 1.8 \times 10^{-5}$ kg/ms, $\bar{V}_{face} = 0.5$ m/s, $d_{fb} = 400 \times 10^{-6}$ m, $d_p = 15 \times 10^{-6}$ m, $d_{fb-fb} = 100 \times 10^{-6}$ m.

Find $\eta_{c,15\ \mu m}$, $d_{p,cut}$.

(a)

$$Stk = \frac{d_p^2 \rho_p}{18 \mu_g} \times \frac{\bar{V}_{face}}{d_{fb}} = \frac{(15 \times 10^{-6})^2 \times 1800}{18 \times 1.8 \times 10^{-5}} \times \frac{0.5}{400 \times 10^{-6}} = 1.563$$

$$\eta_{fb,15\ \mu m} = [Stk/(Stk + 0.85)]^{2.2} = (1.563/2.413)^{2.2} = 0.385$$

Fraction blocked area $= 400/(400 + 100) = 0.8$.

Then:

$$\eta_{15\ \mu m} = \%\ blocked \times \eta_{fb,15\ \mu m} = 0.8 \times 0.385 = 0.308 = 30.8\%$$

(b) For the cut diameter:

$$\eta_{cut} = \%\ blocked \times \eta_{f,\ cut}$$

$$0.5 = 0.8 \times \eta_{fb},\ i.e.\ \eta_{fb,\ cut} = 0.625$$

$$\eta_{f,\ cut} = [Stk/(Stk + 0.85)]^{2.2}$$

$$0.625 = [Stk/(Stk + 0.85)]^{2.2}\ i.e.\ Stk = 3.6$$

Then:

$$Stk = \frac{d_p^2 \rho_p}{18 \mu_g} \times \frac{\bar{V}_{face}}{d_{fb}}$$

$$3.6 = \frac{d_{p,cut}^2 \times 1800}{18 \times 1.8 \times 10^{-5}} \times \frac{0.5}{400 \times 10^{-6}}$$

$$d_{p,\ cut} = 22.8 \times 10^{-6}\ m = 23\ \mu m$$

(c) A two-layer filter can be modelled as two collectors (1, 2) of equal collection efficiency in series:

$$\eta_{overall,\ 15\ \mu m} = 1 - [(1 - \eta_1)(1 - \eta_2)] = 1 - (1 - \eta_{single\ row,\ 15\ \mu m})^2$$

$$= 1 - (1 - 0.308)^2 = 0.521 = 52.1\%$$

Worked Example 5.3 – Cyclone separator performance

(a) A gas stream containing particles with a density of 1800 kg/m³ is passed through a cyclone separator with an overall diameter of 1 m at a rate of 1 kg/s.
   If the gas temperature is 25 °C at a pressure of 1 bar (abs), using the data below, determine:
   (i) The number of effective turns (N) of the gas stream.
   (ii) The collection efficiency for 10 μm particles.

Data
   Specific gas constant: 287 J/kg K, gas dynamic viscosity: 1.8×10^{-5} kg/ms.
(b) A consultant claims that the overall collection efficiency can be raised to *above 90%* by making *EITHER* of the following individual modifications to the separator operating parameters:
   (iii) Increase the length of the cyclone cone (H_2) by 0.5 m.
   (iv) Increase the temperature of the gas to 50 °C.
   Is the consultant correct in *each individual* case? Quantify your answer.
   Note that the following may be used in part (b)iv:

$$\mu_{\text{gas stream}} \ (\text{kg/ms}) \ \text{at} \ T(\text{Kelvin}) = 1.71 \times 10^{-5} \times (T(\text{Kelvin})/273)^{0.7}$$

Solution
Given: $\rho_p = 1800$ kg/m³, $D_0 = 1$ m, $\mu_g = 1.8 \times 10^{-5}$ kg/ms,

$$R = 287 \ \text{J/kg K}, \ P = 1 \ \text{bar} = 1 \times 10^5 \ \text{Pa},$$
$$T = 25 °C = 298 \ \text{K}, \ d_p = 10 \times 10^{-6} \ \text{m}, \ \dot{m} = 1 \ \text{kg/s}.$$

Find (a) N, $\eta_{C, \ 10 \ \mu m}$.
   (b) New $\eta_{C, \ 10 \ \mu m}$ with parameter changes.
From *'classical'* cyclone dimensions:

$$D_o = 1 \ \text{m}, \ D_e = 0.5 \ \text{m}, \ r_2 = 0.5 \ \text{m}, \ r_1 = 0.25 \ \text{m}, \ H = 0.5 \ \text{m}, \ H_1 = 2 \ \text{m}, \ H_2 = 2 \ \text{m}$$

$$\rho_g = P/RT$$
$$= 1 \times 10^5 / (287 \times 298) = 1.17 \ \text{kg/m}^3$$

$$\dot{m} = \rho \dot{V}$$
$$1 = 1.17 \dot{V}$$
$$\dot{V} = 0.854 \ \text{m}^3/\text{s}$$

$$C_2 = \frac{4r_2^3}{(r_2 + r_1)(r_2^2 - r_1^2)}$$
$$= 4 \times 0.5^3 / [(0.5 + 0.25) \times (0.5^2 - 0.25^2)] = 3.55$$

(a)i

$$N = (1/H)(2H_1 + H_2)$$
$$= (1/0.5)((2 \times 2) + 2) = 12 \ \text{turns}$$

(a)ii

$$Stk_{ave} = \left(\frac{\rho_p d_p^2}{18 \mu_g}\right) \times \left(\frac{\dot{V}}{r_2 H (r_2 - r_1)}\right)$$

$$= [1800 \times (10 \times 10^{-6})^2 / 18 \times 1.8 \times 10^{-5}] \times [0.854/(0.5 \times 0.5(0.25))]$$

$$= 0.00759$$

$$\eta_c = 1 - \exp[-Stk_{ave} \times C_2 \times 2\pi N]$$

$$= 1 - \exp[-0.00759 \times 3.55 \times 2 \times \pi \times 12] = 0.869, \text{ i.e. } 86.9\%$$

(b)iii Increase H_2 to 2.5 m

$$N = (1/H)(2H_1 + H_2)$$

$$= (1/0.5)((2 \times 2) + 2.5) = 13 \text{ turns}$$

$$\eta_c = 1 - \exp[-Stk_{ave} \times C_2 \times 2\pi N]$$

$$= 1 - \exp[-0.00759 \times 3.55 \times 2 \times \pi \times 13] = 0.889, \text{ i.e. } 88.9\%.$$

(b)iv Increase the gas temperature to 50 °C

$$T = 50\,°C = 323 \text{ K}$$

$$\rho_g = P/RT = 1 \times 10^5/(287 \times 323) = 1.078 \text{ kg/m}^3$$

$$\dot{V} = \dot{m}/\rho_g = 1/1.078 = 0.927 \text{m}^3/\text{s}$$

$$\mu_g = 1.71 \times 10^{-5} \times (T/273)^{0.7}$$

$$= 1.71 \times 10^{-5} \times (323/273)^{0.7} = 1.923 \times 10^{-5} \text{kg/ms}$$

$$Stk_{ave} = \left(\frac{\rho_p d_p^2}{18 \mu_g}\right) \times \left(\frac{\dot{V}}{r_2 H (r_2 - r_1)}\right) = 7.71 \times 10^{-3}$$

$$\eta_c = 1 - \exp[-Stk_{ave} \times C_2 \times 2\pi N]$$

$$= 1 - \exp[-0.00771 \times 3.55 \times 2 \times \pi \times 12] = 0.873, \text{ i.e. } 87.3\%.$$

The suggested improvements are both unsubstantiated.

5.6 Tutorial Problems

5.1 A pollution control device has the fractional efficiency profile shown below. Determine its total collection efficiency (%) when attempting to clean up a gas stream contaminated with particulates having the weight fractions indicated.

Particle size (μm)	Fractional efficiency (%)	Weight fraction (%)
50	99	10
30	95	30
10	85	35
5	60	15
1	55	10

[Answer: 82.65%]

5.2 A polluted air stream has its particulate concentration reduced from 175 μg/m^3 to 10 μg/m^3 by a collection device.

Assuming the volume flow rate remains unchanged, determine the collection efficiency, penetration and decontamination factor for the collection device.

[Answers: 94.29%, 5.71, 17.5]

5.3 A filter has a collection efficiency of 80%.

Determine the overall collection efficiency (%) of three such filters used in series.

[Answer: 99.2%]

5.4 Two pollution control devices are connected in parallel.

The collection efficiencies and flow rate split are shown below.

Collector	Collection efficiency (%)	Flow rate split (%)
1	85	55
2	80	45

Determine the overall collection efficiency (%) for the system.

[Answer: 82.75%]

5.5 A gravity settler with a length of 10 m, a width of 5 m and a height of 2 m processes 600 m^3/min of particulate-laden gas.

Compute, using a *well-mixed* model, the collection efficiency (%) for this unit at a particle size of 40 μm.

Take $\rho_p = 2000$ kg/m^3 and $\mu_g = 1.8 \times 10^{-5}$ kg/ms.

A consultant claims that the overall collection efficiency can be raised to over 50% by making *any one* of the following individual modifications to the settler operating parameters:

- Increase the height of the settler by 50%.
- Increase the settler length by 2 m.
- Decrease the temperature of the gas to 10 °C.

Is the consultant correct *in each individual case*? Quantify your answer.

Note that the following property correlation may be used:

$$\mu_{\text{gas stream}} \text{ (kg/ms) at } T(\text{Kelvin}) = 1.71 \times 10^{-5} \times (T(\text{Kelvin})/273)^{0.7}$$

[Answers: 38.4%, No (38.4%), No (44%), No (39.1%)]

5.6 A particulate gas enters a cyclone separator with an inlet velocity of 8 m/s. If the device processes gas at a rate of 0.2 m³/s, determine the dimensions (m) of the cyclone assuming 'classical' proportions.

[Answer: $W_i = 0.112$ m, $D_o = 0.448$ m, $H = 0.224$ m etc.]

5.7 A 50 m³/s gas stream containing 2 μm diameter particles passes through an ESP of electric field strength 300 kV/m.
If the plate surface area is 3500 m², using a well-mixed model determine:
(a) The particle drift velocity (m/s).
(b) The collection efficiency (%).
(c) The ESP total plate area (m²) required for a collection efficiency of 99.9%.
Take the gas stream dynamic viscosity as 1.8×10^{-5} kg/ms, the particle density as 2000 kg/m³ and the particle dielectric constant as 6.

[Answers: 0.066 m/s, 99%, 5233 m²]

5.8 An ESP is treating a particle-laden gas stream with a collection efficiency of 98%. If all other operating parameters remain constant, determine the new collection efficiency (%) if the gas stream flow rate is increased by 50%.

[Answer: 92.63%]

5.9 A manufacturer proposes to produce filters with the following specification for the collection of 1 μm particles in an air stream:
Air viscosity: 1.8×10^{-5} kg/ms Particle density: 2000 kg/m³
Filter thickness: 3 mm Filter face velocity: 0.3 m/s
Filter fibre solid fraction: 5% Cylindrical fibre diameter: 10 μm
Show that this design is likely to have a poor collection efficiency.
It is required to raise the collection efficiency to 95%.
Keeping all other parameters unchanged, the options are:
(a) Increasing the fibre solid fraction (f_{fibre}).
(b) Decreasing the fibre diameter (d_{fibre}).
Calculate the new parameter value *in both cases separately*.

[Answers: 36.5%, 25.76%, 4.9×10^{-6} m]

5.10 A 5 m high counter-flow spray chamber uses 1 mm diameter water droplets as the scrubbing liquid to capture particles with a velocity of 3 m/s in a contaminated gas stream with a temperature of 50 °C.
If the particle size is 10 μm, using the data below determine the overall collection efficiency for the chamber when the flow rate ratio of gas to liquid is 5000.

Data
Gas dynamic viscosity at 50 °C: 1.95×10^{-5} kg/ms
Particle density: 2000 kg/m³, water density: 1000 kg/m³.
A consultant claims that the overall collection efficiency can be raised to over 80% by making *any one* of the following individual modifications to the chamber operating parameters:
• Increase the height of the chamber by 1 m.

- Increase the liquid flow rate by 10%.
- Decrease the temperature of the gas to 10 °C.

Is the consultant correct in *each individual* case? Quantify your answer. Note that the following property correlation may be used:

$$\mu_{gas\ stream}\ (kg/ms)\ at\ T(Kelvin) = 1.71 \times 10^{-5} \times (T(Kelvin)/273)^{0.7}$$

[Answers: 78.5%, Yes (84.1%), Yes (87.6%), Yes (84.3%)]

6

Carbon Capture and Storage

6.1 Overview

Carbon-based fuels, i.e. coal, crude oil and natural gas or their derivatives, underpin human civilization at present.

The complete combustion of carbon with oxygen results in carbon dioxide. This is simple chemistry.

Carbon dioxide is one of many climate-modifying (or *greenhouse*) gases that have been released routinely to the atmosphere in ever-increasing amounts for several centuries.

It has been estimated by the Intergovernmental Panel on Climate Change (IPCC) that, since 1750, the cumulative CO_2 emissions to the atmosphere from human development amount to approximately 2000 Gt CO_2, with around 1000 Gt CO_2 being added in the last 40 years alone. It is now understood that these gases affect the energy interchange with the planet's solar input, altering its heat balance in favour of warmer average conditions. Current modelling predicts that the results of this warming are likely to be unwelcome from a human perspective. The global current (2016) CO_2 emission rate is approximately 40 Gtonnes per annum.

For the future, a translation of energy usage away from fossil fuels and into low/no carbon sources, for example, renewable technologies, must be effected. However, it is not practical to make this change overnight and, in the interim, technologies such as carbon capture and storage must be developed to reduce the impact of these emissions.

Learning Outcomes

- To understand the thermo-physical properties of carbon dioxide.
- To understand the behaviour of gas mixtures incorporating carbon dioxide.
- To predict the minimum thermodynamic energy requirement in gas mixture separation.
- To gain an awareness of gas separation methods such as chemical absorption, physical absorption and membrane technology.
- To be able to provide an overview of the operational aspects of dealing with the transport of separated carbon dioxide, for example, gas compression, pipework design and HSE issues.
- To gain a basic knowledge of carbon storage alternatives.

Conventional and Alternative Power Generation: Thermodynamics, Mitigation and Sustainability,
First Edition. Neil Packer and Tarik Al-Shemmeri.
© 2018 John Wiley & Sons Ltd. Published 2018 by John Wiley & Sons Ltd.

6.2 Thermodynamic Properties of CO_2

6.2.1 General Properties

The formation of carbon dioxide during combustion is a result of a chemical combination of carbon with oxygen. Some basic facts about the reactants and product are given below:

Oxygen (O):

Is a non-metal, colourless, odourless gas at normal temperature and pressure.

Is the third most abundant element in the universe.

Has a relative atomic mass of 15.9994, usually truncated to 16.

Has a melting point of $-219\,°C$ and a boiling point of $-183\,°C$.

Comprises approximately 50% of the earth's crust and $> 85\%$ of the oceanic mass.

Is highly reactive, forming oxides with other elements.

Has a diatomic gaseous form (O_2), which comprises approximately 21% by volume of the atmosphere.

Has a molar mass of 16 g/mol.

Carbon (C):

Is a solid non-metal.

Exists in three forms or allotropes: clear, cubic crystal *diamond*, hexagonal black crystal *graphite* and fullerene (nanotubes/nanospheres).

Has a relative atomic mass of 12.0107, usually truncated to 12.

Is a tetravalent atom, i.e. it can combine with up to four other atoms. At the time of writing, about 20 million carbon compounds are known to be in existence.

Sublimates before melting in its solid form. Its sublimation point is $3642\,°C$.

Has a molar mass of 12 g/mol and is the 15th most common element in the earth's crust.

Carbon dioxide (CO_2):

Is a colourless, odourless, non-combustible gas at normal temperature and pressure.

Comprises two oxygen atoms in combination with a single carbon atom. The bonds between the carbon and oxygen atoms are double in nature.

Is heavier than air.

Has a molecular mass of 44.01, usually truncated to 44, giving a molar mass of 44 g/mol.

Is a natural component of air, with current concentrations of between 380 and 400 ppm_v.

Has doubled in concentration since the start of the Industrial Revolution.

Has a recommended permissible human exposure limit of 5000 ppm_v.

Produces a slightly acidic solution when dissolved in water (carbonic acid).

Is important in many biological processes:

It is essential to plant development, being extracted from ambient air by the process of photosynthesis.

It is excreted by living organisms via respiration.

It is also released into the environment by the decomposition of living matter.

Carbon dioxide is generated in large quantities by several man-made processes including the cement and ammonia production industries.

Carbon dioxide has many industrial uses and applications in the food, pharmaceutical and healthcare industries, performing cooling, freezing and process temperature control, acidification/neutralization, carbonization and transportation duties.

It is also found in large quantities associated with natural gas deposits and can be injected into old oil wells to help maintain crude oil production by partial repressurization and the lowering of the oil viscosity.

Most importantly, it is generated as a by-product in the combustion of hydrocarbon ('fossil') fuels and is an atmospheric climate-modifying gas.

Collections of carbon dioxide's thermodynamic properties like pressure, volume, temperature, enthalpy and entropy are commonly determined using property tables or 2-D and 3-D graphs and charts. For example, the pressure–temperature relationship for carbon dioxide is shown in Figure 6.1.

In the figure, the lines illustrate the set of conditions (pressure, temperature) under which phase change occurs.

Crossing the line from a liquid to a solid condition is termed *fusion*. The reverse action is termed *melting*.

Crossing the line from a liquid to a gas condition is termed *vaporization*. The reverse action is termed *condensation*.

Crossing the line from a solid to a gas condition is termed *sublimation*. The reverse action is termed *deposition*.

The *triple point* indicates conditions under which three phases can coexist in equilibrium. Triple point conditions for carbon dioxide are approximately −56 °C and 5.1 bar.

The *critical point* is a set of conditions under which the saturated liquid and saturated vapour states are identical. This point is indicated on all three main property relationship diagrams, i.e. $P–v$, $P–T$ and $T–v$.

On a temperature–volume ($T–v$) basis, the critical point indicates that pressure above which phase change occurs without any saturated liquid–vapour (or mixed-phase) region.

On a pressure–volume ($P–v$) basis, the critical point indicates that temperature above which phase change occurs without any saturated liquid–vapour (or mixed-phase) region.

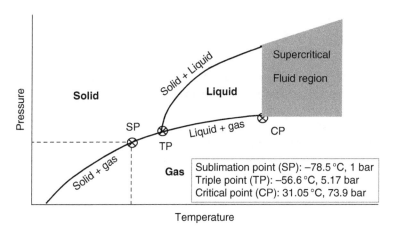

Figure 6.1 Pressure–temperature ($P–T$) property diagram for carbon dioxide.

Critical point conditions for carbon dioxide are 304.2 K (T_{cr}) and 73.9 bar (P_{cr}). The specific volume at the critical point (v_{cr}) is approximately 2.137×10^{-3} m^3/kg. Substances existing beyond critical point conditions are termed *supercritical*.

Supercritical fluids are useful in that they possess the density of the fluid phase whilst maintaining the viscosity of the gas phase.

6.2.2 Equations of State

Expressions (or equations) of state are available for use in approximating the relationship between basic properties, for example, pressure, temperature and volume.

Equations for predicting solid and liquid properties are quite complex, with many coefficients. Some of the more common expressions for gases are presented below.

6.2.2.1 The Ideal or Perfect Gas Law

This is a suitable approximation for high-temperature and low-pressure gaseous conditions. It is expressed in several forms, as shown in Table 6.1.

Near the saturation curve, the critical point behaviour tends to deviate from the ideal, and this can be accounted for by the use of a *compressibility factor, z* or other predictive formulae.

6.2.2.2 The Compressibility Factor

This is expressed as the ratio of the actual specific volume divided by the ideal specific volume, i.e. $z = v_{act}/v_{ideal}$.

The ideal gas law relationship (on a specific volume basis) is then modified to:

$$Pv = zRT \tag{6.1}$$

The compressibility factor is usually determined by the use of an empirically derived *compressibility chart* (see Figure 6.2), where the prevailing pressure and temperatures are normalized with their critical point values to produce so-called *reduced* values, i.e.

Reduced pressure P_r (Pa) $= P/P_{cr}$
Reduced temperature T_r (K) $= T/T_{cr}$
Pseudo reduced volume v_r (m^3/kg) $= v/v_{cr} = vP_{cr}/RT_{cr}$

6.2.2.3 Van der Waal Equation of State

A third equation of state was proposed by van der Waal in 1873. This takes into account the fact that the molecules of the gas have volume themselves and are not point sources.

Table 6.1 Forms of the ideal gas law.

Basis	Expression of ideal gas law
Mass (m, kg)	$PV = mRT$
Molar (n, kmol)	$PV = nR_oT$
Mass independent (v, m^3/kg)	$Pv = RT$
Molar independent ($\bar{v}$, m^3/kmol)	$P\bar{v} = R_oT$

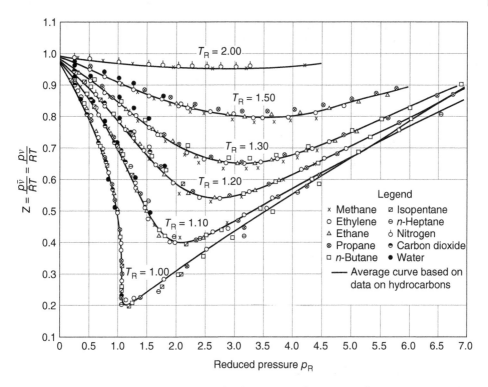

Reduced pressure p_R

Figure 6.2 Compressibility chart. Reproduced with permission from Moran, Shapiro, Boettner and Bailey (2012) *Principles of Engineering Thermodynamics*, 7th edition, John Wiley & Sons.

It also acknowledges the existence of inter-molecular forces and non-elastic collisions, thus:

$$[P + (a/v^2)](v - b) = RT \tag{6.2}$$

where a, b are factors again based on the critical point values for the gas:

$$a = \frac{27R^2 T_{cr}^2}{64P_{cr}}$$

$$b = \frac{RT_{cr}}{8P_{cr}}$$

6.2.2.4 Beattie–Bridgeman Equation (1928)

This improvement uses the molar-independent volume ($\bar{v}$, m³/mol).

$$P = \frac{R_o T}{\bar{v}^2}\left(1 - \frac{c}{\bar{v}T^3}\right)(\bar{v} - B) - \frac{A}{\bar{v}^2} \tag{6.3}$$

where:

$$A = A_o\left(1 - \frac{a}{\bar{v}}\right) \text{ and } B = B_o\left(1 - \frac{b}{\bar{v}}\right)$$

If pressure is in kPa, temperature in Kelvin, universal gas constant $R_o = 8.314$ kPa.m^3/kmol and molar specific volume $\bar{v}$ is in m^3/kmol then the five constants for carbon dioxide are:

$$A_o = 507.283, \quad a = 0.07132, \quad B_o = 0.10476, \quad b = 0.07235, \quad c = 6.6 \times 10^5$$

This expression has good accuracy for $\rho \leq 0.8\rho_{cr}$.

6.2.2.5 Benedict–Webb–Rubin Equation (1940)

This has improved accuracy by virtue of an increased number of constants.

$$P = \frac{R_o T}{\bar{v}} + \left(B_o R_o T - A_o - \frac{C_o}{T^2}\right)\frac{1}{\bar{v}^2} + \frac{bR_o T - a}{\bar{v}^3} + \frac{a\alpha}{\bar{v}^6} + \frac{c}{\bar{v}^3 T^2}\left(1 + \frac{\gamma}{\bar{v}^2}\right)e^{-\gamma/\bar{v}^2}$$

(6.4)

If pressure is in kPa, temperature in Kelvin, universal gas constant $R_o = 8.314$ kPa.m^3/kmol and molar specific volume $\bar{v}$ is in m^3/kmol then the eight constants for carbon dioxide are:

$a = 13.86$, $A_o = 277.3$, $b = 0.007210$, $B_o = 0.04991$, $c = 1.511 \times 10^6$, $C_o = 1.404 \times 10^7$, $\alpha = 8.470 \times 10^{-5}$, $\gamma = 0.00539$

This expression has good accuracy for $\rho \leq 2.5\rho_{cr}$.

6.2.2.6 Peng–Robinson Equation of State (1976)

This is proposed as a model for both gaseous *and* liquid phases.

$$P = \frac{RT}{\bar{v} - b} - \frac{a\alpha}{\bar{v}^2 + 2b\bar{v} - b^2}$$

(6.5)

where

$$a = \frac{0.457235R^2 T_{cr}^2}{P_{cr}}$$

$$b = \frac{0.077796RT_{cr}}{P_{cr}}$$

$$\alpha = (1 + K(1 - T_r^{0.5}))^2$$

$$K = 0.37464 + 1.54226\omega - 0.26992\omega^2$$

For carbon dioxide, ω (the *acentric factor*) has a value of approximately 0.225.

6.3 Gas Mixtures

Carbon dioxide is commonly associated in a mixture with other gases, for example, the atmosphere, products of combustion etc., rendering an understanding of gas mixtures important.

The analysis of gas mixtures is grounded on component composition. Compositional descriptions of mixtures can be either mole- or mass-based.

6.3.1 Fundamental Mixture Laws

Consider a mixture comprising a number of gases.

From a matter conservation perspective, the mixture's total mass must be equal to the sum of the masses of its individual components. Similarly for its molar components.

The contribution of any individual component (i) to the total can be described by either a mass fraction (mf_i) or a molar fraction (y_i). Note that in both mass and molar cases, the fractional sum is equal to unity.

If the gas mixture comprises j components in total, the above considerations can be more formally stated as shown in Table 6.2.

Mass fraction analyses are often referred to as *gravimetric* whilst molar fraction analyses are termed *volumetric*.

The relationship between the number of kilomoles of a substance (n, kmol), its mass (m, kg) and molar mass (M, kg/kmol) is given by:

$$m = nM \tag{6.6}$$

Knowledge of the *mixture's* molar mass (M_{mix}) is important, as it helps to determine the *mixture's* gas constant (R_{mix}) and define the relationship between mass and molar fractions. Rearranging the mass-to-mole converter and applying to a mixture:

$$M_{mix} = \frac{m_{mix}}{n_{mix}} = \frac{\sum m_i}{n_{mix}} = \frac{\sum n_i M_i}{n_{mix}} = \sum_{i=1}^{j} y_i M_i \tag{6.7}$$

Remembering that $R = R_o/M$ then:

$$R_{mix} = \frac{R_o}{M_{mix}} \tag{6.8}$$

Again, the mass–molar converter can be used to determine the fractional relationship for any gas component (i):

$$mf_i = \frac{m_i}{m_{mix}} = \frac{n_i M_i}{n_{mix} M_{mix}} = y_i \frac{M_i}{M_{mix}} \tag{6.9}$$

6.3.2 PVT Behaviour of Gas Mixtures

Ideal gas conditions assume that molecules have negligible volume and do not interact with one another. The simplest gas mixture models assume that these conditions hold good for all components of the mixture.

In this way, the contribution that component gases have on the physical behaviour of a mixture can be approximated by using two important laws: *Dalton's Law* and *Amagat's Law*.

Table 6.2 Unit comparison of fundamental mixture laws.

Parameter	Mass basis	Molar basis
Totals	$m_{mix} = \sum_{i=1}^{j} m_i$	$n_{mix} = \sum_{i=1}^{j} n_i$
Fraction	$mf_i = \dfrac{m_i}{m_{mix}}$	$y_i = \dfrac{n_i}{n_{mix}}$

6.3.2.1 Dalton's Law

Consider two gases A and B at the same temperature and volume but at different pressures P_A and P_B, respectively (see Figure 6.3). If these gases are now mixed and the mixture exists at the same temperature and volume as its components, then Dalton's Law says that the resulting pressure of the mixture will be the sum of P_A and P_B. In short:

$$P_{mix} = \sum P_i \text{ at } T_{mix}, V_{mix} \tag{6.10}$$

The ratio of any individual pressure of a component gas to the total gas mixture pressure (P_i/P_{mix}) is called the *pressure fraction*. Using the ideal gas law, it can be shown that:

$$\frac{P_i}{P_{mix}} = y_i \tag{6.11}$$

That is, the pressure fraction of any gas component in a mixture of gases is equal to the molar fraction of the component gas in the mixture.

The product of the molar fraction and the mixture pressure $(y_i P_{mix})$ is called the *partial pressure*.

6.3.2.2 Amagat's Law

Consider two gases A and B at the same temperature and pressure but having different volumes V_A and V_B, respectively (see Figure 6.4). If these gases are now mixed and the mixture exists at the same temperature and pressure as its components, then Amagat's Law says that the resulting volume of the mixture will be the sum of V_A and V_B. In short:

$$V_{mix} = \sum V_i \text{ at } T_{mix}, P_{mix} \tag{6.12}$$

The ratio of any individual volume of a component gas to the total gas mixture volume (V_i/V_{mix}) is called the *volume fraction*. Again, using the ideal gas law, it can be shown that:

$$\frac{V_i}{V_{mix}} = y_i \tag{6.13}$$

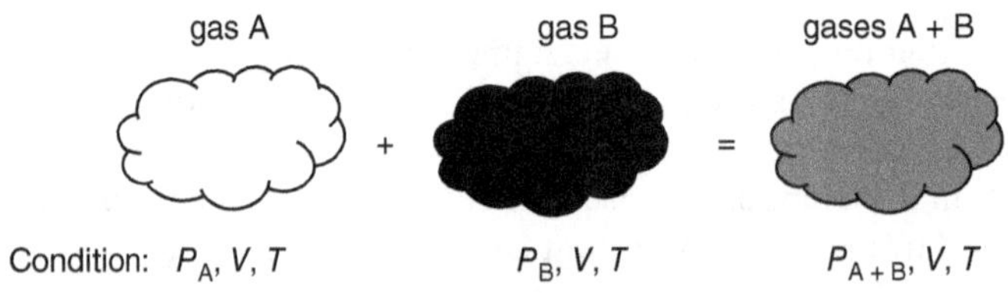

gas A   gas B   gases A + B

Condition: P_A, V, T P_B, V, T P_{A+B}, V, T

Figure 6.3 Illustration of Dalton's Law.

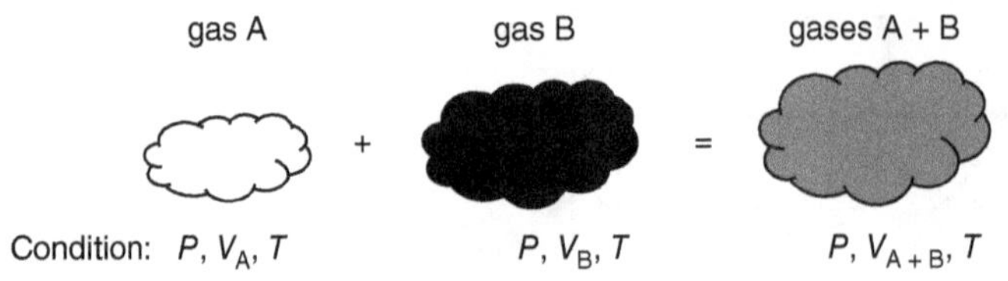

gas A   gas B   gases A + B

Condition: P, V_A, T   P, V_B, T   P, V_{A+B}, T

Figure 6.4 Illustration of Amagat's Law.

That is, the volume fraction of a gas component in a mixture of gases is equal to the molar fraction of the component gas in the mixture.

The product of the molar fraction and the mixture volume ($y_i V_{mix}$) is called the *partial volume.*

Real gases can be analyzed using Amagat's Law providing compressibility is taken into account.

The compressibility factor of a gas mixture (z_{mix}) can be evaluated by summing the product of the molar fraction and the compressibility factor for each individual gas component:

$$z_{mix} = \sum_{i=1}^{j} y_i z_i \tag{6.14}$$

A second approach uses *Kay's rule,* which regards the mixture as a *pseudopure* substance. This introduces the concepts of a pseudocritical pressure (P^*_{cr}) and pseudocritical temperature (T^*_{cr}) where:

$$P^*_{cr} = \sum_{i=1}^{j} y_i P_{cr,i} \quad \text{and} \quad T^*_{cr} = \sum_{i=1}^{j} y_i T_{cr,i} \tag{6.15}$$

Alternatively, use can be made of van der Waals, Beattie–Bridgeman or the Benedict–Webb–Rubin equations, basing the equation coefficients on the coefficients of the mixture components.

6.3.3 Thermodynamic Properties of Gas Mixtures

The approach to mixture property evaluation is based on an extension to Dalton's Law called the *Gibbs–Dalton Law.* This assumes that each component gas has a thermodynamic property (i.e. enthalpy, internal energy, entropy) value equal to the value it would have if it occupied the mixture volume alone at the same temperature.

Mass or molar properties of mixtures can be determined by *addition* of component contributions (i).

Let ψ be any *mass-independent* thermodynamic property and $\bar{\psi}$ be its molar-independent counterpart, then mixture property values can be evaluated as shown in Table 6.3.

For example:

$$s_{mix} = \sum_{i=1}^{j} mf_i s_i \quad \text{and} \quad \bar{s}_{mix} = \sum_{i=1}^{j} y_i \bar{s}_i$$

Table 6.3 Comparison of gravimetric and volumetric analyses.

For gravimetric analyses (mass fraction, *mf*)	For volumetric analyses (molar fraction, *y*)
$\psi_{mix} = \sum_{i=1}^{j} mf_i \psi_i$	$\bar{\psi}_{mix} = \sum_{i=1}^{j} y_i \bar{\psi}_i$

Table 6.4 Unit comparison of mixture thermodynamic properties.

In general	Mass based (m, kg)	Molar based (n, kmol)
$\displaystyle \Psi_{\text{mix}} = \sum_{i=1}^{j} \Psi_i$	$\displaystyle \Psi_{\text{mix}} = \sum_{i=1}^{j} m_i \psi_i$	$\displaystyle \Psi_{\text{mix}} = \sum_{i=1}^{j} n_i \bar{\psi}_i$

Let ψ be any mass- or molar-*dependent* thermodynamic property, for example, U, H, S etc., then mixture property values can be evaluated as shown in Table 6.4.

For example:

$$S_{\text{mix}} = \sum_{i=1}^{j} S_i = \sum_{i=1}^{j} m_i s_i = \sum_{i=1}^{j} n_i \bar{s}_i$$

The above relationships are exact for ideal gases and approximate for real gases.

Property changes in a mixture are evaluated by summing the individual component changes. For example:

$$\Delta S_{\text{mix}} = \sum_{i=1}^{j} \Delta S_i = \sum_{i=1}^{j} m_i \Delta s_i = \sum_{i=1}^{j} n_i \Delta \bar{s}_i$$

Evaluation of enthalpy and internal energy changes only requires knowledge of the initial and final temperature. However, entropy changes are both temperature and pressure sensitive, thus, for any component:

$$\Delta s_i = C_{\text{p},i} \ln \frac{T_{i,2}}{T_{i,1}} - R_i \ln \frac{P_{i,2}}{P_{i,1}} \quad \text{and}$$

$$\Delta \bar{s}_i = \bar{C}_{\text{p},i} \ln \frac{T_{i,2}}{T_{i,1}} - R_\text{o} \ln \frac{P_{i,2}}{P_{i,1}} \tag{6.16}$$

assuming a constant heat capacity value.

Consider two gases at the same initial temperature and pressure. If, after mixing, the mixture retains the same temperature and pressure, the associated entropy change is evaluated as follows:

$$\Delta S_{\text{mix}} = \sum_{i=1}^{j} n_i \Delta \bar{s}_i$$

$$= \sum_{i=1}^{j} n_i \left[\bar{C}_{\text{p},i} \ln \frac{T_{i,2}}{T_{i,1}} - R_\text{o} \ln \frac{P_{i,2}}{P_{i,1}} \right]$$

Substituting for $P_{i,2}$ using the pressure fraction relationship and acknowledging isobaric conditions, this can be written as:

$$\Delta S_{\text{mix}} = \sum_{i=1}^{j} n_i \left[R_\text{o} \ln y_i \frac{P_{\text{mix}}}{P_{i,1}} \right] = -R_\text{o} \sum n_i \ln y_i \tag{6.17}$$

6.3.4 Thermodynamics of Mixture Separation

In a combustion process, carbon dioxide is generated in combination with other gases, i.e. in a mixture. In order to facilitate its capture and storage, it must first be separated from the other constituents. This will require the expenditure of work. The following analysis approaches the work evaluation from a thermodynamic perspective by first briefly considering the concepts of *exergy, reversible work* and *irreversibility.*

The work potential of a system is evaluated by the property *exergy* (X, kJ). It can be thought of as a measure of a system's thermomechanical state relative to its environment. A system existing at environmental conditions is often said to be in a *dead state.*

Energy is a property that is conserved according to the first law of thermodynamics. However, energy usage can result in the loss of some of the *work potential.*

In the everyday use of energy, exergy can be lost or destroyed by the presence of process irreversibilities such as heat loss, friction etc. Hence, exergy, unlike energy, is not conserved.

Reversible processes take a substance from its initial state to its final state in the absence of *irreversibilities.*

Work taking place in a reversible manner (W_{rev}) will be a maximum for a work production process and a minimum in a process requiring work.

If some useful work is involved, then the exergy destroyed is the difference between the reversible work and the useful work.

If a system changes to its dead state spontaneously, all of its exergy is destroyed.

For work production:

$$X_{destroyed} = W_{rev,out} - W_{useful, out} \tag{6.18}$$

In an irreversible work-generation process, the *useful work* (W_{useful}) produced is less than the reversible work as a consequence of the presence of irreversibilities. Irreversibility can be described in terms of entropy change (ΔS), thus:

$$X_{destroyed} = T_o \Delta S \tag{6.19}$$

where T_o is a reference or environmental temperature.

Mixing of ideal gases is spontaneous and irreversible because work is needed to separate the components.

6.3.4.1 Minimum Separation Work

For an ideal, adiabatic mixing process, the entropy generated is equal to the change in mixture entropy and the exergy destroyed. Using Equations (6.17) and (6.19), the results of this argument are shown in Table 6.5 on molar and molar-independent bases.

Table 6.5 Unit comparison of mixture entropy change and exergy destruction.

Basis	Mixture entropy change	Units	Exergy destroyed in mixing	Units
Molar	$\Delta S_{mix} = -R_o \sum_i n_i \ln y_i$	kJ/K	$X_{destroyed} = -R_o T_o \sum_i n_i \ln y_i$	kJ
Molar independent	$\Delta \bar{s}_{mix} = -R_o \sum_i y_i \ln y_i$	kJ/kmol K	$\bar{x}_{destroyed} = -R_o T_o \sum_i y_i \ln y_i$	kJ/kmol

Table 6.6 Comparison of total separation work expressions.

Basis	Work expression	Units
Molar independent (y_i)	$\bar{w}_{\text{min,in}} = -R_o T_o \sum_i y_i \ln y_i$	kJ/kmol
Molar (n_i)	$W_{\text{min,sep}} = -R_o T_o \sum_i n_i \ln y_i$	kJ
Molar flow ($\dot{n}_{\text{mix}}$)	$\dot{W}_{\text{min,sep}} = -\dot{n}_{\text{mix}} R_o T_o \sum_i y_i \ln y_i$	kW
Mass flow ($\dot{m}$)	$\dot{W}_{\text{min,sep}} = -\dot{m}_{\text{mix}} R T_o \sum_i y_i \ln y_i$	kW

For an ideal mixing process with no useful work generated, i.e. in the absence of irreversibilities, the reversible work is equal to the exergy destroyed. Applying the principle of reversibility, this must also be the minimum work input ($W_{\text{min,sep}}$) for a mixture separation process.

The minimum work of separation can be expressed in molar fraction, molar, molar flow ($\dot{n}_{\text{mix}}$, kmol/s) or mass flow ($\dot{m}_{\text{mix}}$, kg/s) terms, with corresponding units as shown in Table 6.6.

Note that the mass-flow-based expression in Table 6.6 was obtained by using the mass-to-mole converter, thus:

$$\dot{m}_{\text{mix}} = \dot{n}_{\text{mix}} M_{\text{mix}} \tag{6.20}$$

The above is applicable for the complete separation of components.

If separation is incomplete, the minimum work is determined by calculating the difference between incoming and outgoing mixtures.

6.3.4.2 Separation of a Two-component Mixture

Let a gas mixture comprising two components A and B contain n_A kmols of A and n_B kmols of B. The corresponding mole fractions are y_A and y_B.

The minimum work of separation can be expressed in either mass or molar terms, with corresponding units as shown in Table 6.7.

If the removal of one component (A) of a large mixture leaves the remaining mixture relatively unchanged, the minimum work required to remove A (kJ/kmol of A) is in

Table 6.7 Comparison of two-component separation work expressions.

Basis	Work expression	Units
Molar independent (y_i)	$\bar{w}_{\text{min,sep}} = -R_o T_o (y_A \ln y_A + y_B \ln y_B)$	kJ/kmol of A
Molar (n_i)	$W_{\text{min,sep}} = -R_o T_o (n_A \ln y_A + n_B \ln y_B)$	kJ
Molar flow ($\dot{n}_{\text{mix}}$)	$\dot{W}_{\text{min,sep}} = -\dot{n}_{\text{mix}} R_o T_o (y_A \ln y_A + y_B \ln y_B)$	kW
Mass flow ($\dot{m}$)	$\dot{W}_{\text{min,sep}} = -\dot{m}_{\text{mix}} R T_o (y_A \ln y_A + y_B \ln y_B)$	kW

molar terms, given by:

$$\bar{w}_{\text{min,sep}} = -R_o T_o \ln y_A = R_o T_o \ln(1/y_A) \tag{6.21}$$

and in mass-independent terms (kJ/kg of A):

$$W_{\text{min,sep}} = -R_A T_o \ln y_A = R_A T_o \ln(1/y_A) \tag{6.22}$$

The energy requirement will, of course, change as the removal proceeds because of the changes in mass and partial pressure.

A mass-based estimate of the average energy requirement can be made by employing numerical integration (trapezoidal or the Simpson rule) between the process's start and end points.

6.4 Gas Separation Methods

6.4.1 Chemical Absorption by Liquids

Chemical absorption of CO_2 is not a new technology. It already has applications in natural gas purification and the production of CO_2 from gas turbine flue gases. These processes are, however, small scale when compared to the requirement of coal-fired power stations.

Carbon dioxide can be removed from a flue gas mixture by washing or 'scrubbing' the mixture with a liquid sorbent or solvent that selectively binds chemically or reacts with CO_2 in an absorber tower.

The gas-loaded liquid is then pumped to a regenerator or 'stripper', where the gas is separated from the gas–liquid mixture by the application of heat.

The energy requirement for the process is solvent specific and is usually expressed as a percentage of the output of the power plant being treated.

The cleansed or *lean* solvent is recycled to the absorber for reuse.

The gas-mixture regeneration process is not 100% efficient and some solvent make-up is required. The take-up of gas by the solvent is dependent on reaction kinetics (temperature, reactant concentrations) and, unlike physical absorption processes, is less reliant on gas partial pressure.

This method is therefore suitable for combustion process flue gas treatment where CO_2 partial pressures are typically < 0.5 MPa.

Water will dissolve CO_2 but, in order to improve reaction rates, substances with properties of both *amines* and *alcohols* dissolved in water are in common use as solvents in chemical absorbents.

Amines are hydrocarbons with one or more of the hydrogen atoms substituted by an ammonia (NH_3) compound.

Alcohols are hydrocarbons with one or more of the hydrogen atoms substituted by a hydroxyl (OH) group.

Commercially available solvents include monoethanolamine (MEA), diethanolamine (DEA) and methyldiethanolamine (MDEA) in aqueous solution. Generically, these are termed *alkanolamines*. Solvent characteristics affecting CO_2 absorption include solvent carbon chain length, the position of the alcohol/alkyl groups on the chain and whether the chain is open or cyclic.

A typical absorber/stripper plant schematic is shown in Figure 6.5.

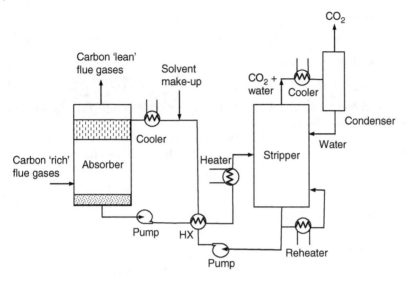

Figure 6.5 Absorber/stripper plant layout schematic.

The flow inputs to the cycle are:

- The combustion flue gases to the absorber
- Make-up solvent.

The flow outputs from the cycle are:

- De-carbonized flue gases from the absorber
- Purified (>99%) CO_2 from the condenser.

For an MEA system, the solvent typically undergoes the following changes:

- Unloaded solvent enters the absorber ($\sim$40 °C) and experiences a small temperature rise ($\sim$15 °C) by exit.
- The (carbon dioxide) loaded solvent leaving the absorber will be heated (by energy exchange with unloaded solvent leaving the stripper and further heated as necessary) before entering the stripper at high temperature ($\sim$100 °C), where the solvent will be separated, leaving a carbon dioxide/water solution.
- The unloaded solvent leaving the stripper ($\sim$90 °C) will be pumped to the absorber (being cooled by heat exchange with loaded solvent leaving the absorber and further cooling as necessary) to then re-enter the absorber to complete its cycle.

The carbon dioxide/water exiting the stripper is dehydrated by a cooler/condenser.

6.4.1.1 Aqueous Carbon Dioxide and Alkanolamine Chemistry

Because of its ionic tendency, water is a good solvent and its hydroxide ion (OH^-) can combine with aqueous carbon dioxide to produce hydrogen carbonate and carbonate ions, thus:

$$CO_2(aq) + OH^- \rightleftarrows HCO_3^-$$

$$HCO_3^- + OH^- \rightleftarrows H_2O + CO_3^{2-}$$

Alkanolamines react with water to form hydroxide ions.
Letting AANH be the generic formula for the alkanolamine:

$$AANH + H_2O \rightleftarrows AANH_2^+ + OH^-$$

Alkanolamines also react with the hydrogen carbonate ion to form carbamate ions, improving reaction rates and reducing equipment size.

$$AANH + HCO_3^- \rightleftarrows H_2O + AANCOO^-$$

6.4.1.2 Alternative Absorber Solutions

Alkali metal (Na/K) carbonates, chilled ammonia and amino acid salts have also being considered as alternative solvents with lower energy regeneration requirements.

The use of *sodium* and *potassium carbonate* solutions (Na_2CO_3, K_2CO_3) pre-dates alkanolamines for CO_2 chemical absorption applications.

The general reaction (for potassium carbonate) is shown below:

$$CO_2 + K_2CO_3 + H_2O \rightleftarrows 2KHCO_3$$

Again, carbonate ions are produced by the reaction.

The sodium/potassium hydrogen carbonate releases carbon dioxide when heated.

These resulting carbonates are again available for further CO_2 take-up. The advantage of using an alkali metal carbonate method is a lower energy requirement. Regeneration energy requirements are, in theory, less than for amine systems. However, having to operate with flue gases at higher pressure conditions (3 bar partial pressure of CO_2 in absorber and *desorption* conditions of 3–4 bar total) would tend to negate this advantage.

Alkali metal methods also have a lower reaction rate and lower solubility. A compromise solution is to use alkali metal carbonates in tandem with DEA.

Ammonia is highly soluble in water:

$$NH_3(aq) + H_2O \rightleftarrows NH_4^+ + HCO_3^-$$

Ammonia/CO_2 chemistry is similar to that of amine/CO_2 in its use of hydrogen carbonate, carbonate and carbamate ion formation.

The formation of carbamate ions is similar to that of alkanolamine solvents, with H_2 replacing the generalized AA component. In general:

$$NH_3 + HCO_3^- \rightleftarrows H_2O + H_2NCOO^-$$

Other ammonia/carbon dioxide/water reactions are manifold, depending on reactant phase as well as temperature, pressure and concentration.

In ammonia-based separation systems, the flue gases first require cleaning and cooling.

The *absorber* operates at temperatures slightly above $0\,°C$ with *desorption* pressures of 15–20 bar. Ammonia recovery is via a wash cycle. Again, regeneration energy requirements are less than for equivalent amine systems.

Amino acid salts are also under consideration, as they possess, to some degree, all the advantages of amine, alkali salt and ammonia systems, in that they are basic, highly soluble in water and are able to form carbamates with CO_2.

6.4.2 Physical Absorption by Liquids

Unlike chemical absorption, physical absorption relies on the ability of a liquid to dissolve a gas and the rate of (concentration-difference-based) diffusion from gas to liquid.

Henry's Law (named in honour of 19th century British chemist William Henry) is concerned with the solubility of a gas in a liquid and states that the mass of any gas (*i*) that will dissolve in a liquid at a constant temperature is directly proportional to the pressure that the gas exerts in contact with the liquid.

For a mixture of gases in contact with the liquid, the pressure of importance is the *partial pressure* contribution of the gas of interest.

The 'law' is, in fact, only an approximate relationship applicable to some gas–liquid solutions at low contaminant gas concentrations and *equilibrium* or *saturation* conditions (signified by *).

Furthermore, the relationship breaks down in the presence of gas–liquid ionization or chemical reaction. The law is commonly expressed thus:

$$P_i = K_H y^*_{i,\,g-l} \tag{6.23}$$

where P_i is the partial pressure of the contaminant gas in the feed gas, $y^*_{i,\,g-l}$ is the mole fraction of the gas in the liquid and K_H is known as *Henry's coefficient* (Pa).

Each gas–liquid system has its own specific value of Henry's coefficient. Again, these values are temperature-dependent. Raising the temperature reduces the liquid solvent's ability to hold the gas.

Dividing the above equation through by the total system pressure (P_{tot}), we have:

$$P_i/P_{tot} = (K_H/P_{tot})y^*_{i,g-l} \tag{6.24}$$

Using Dalton's Law, the left-hand side of the equation becomes the mole fraction of the gas in the gas mixture associated with gas–liquid saturation conditions:

$$y^*_{i,g-g} = (K_H/P_{tot})y^*_{i,g-l} \tag{6.25}$$

This mole-fraction-based expression is most commonly employed to illustrate the relationship in a gas–liquid absorption process. Care must be taken, as values of (K_H/P_{tot}) are sometimes presented in standard texts in contrast to Henry's coefficient, K_H.

Absorber/stripper systems are again used. As a consequence of Henry's Law, systems operate best with high-pressure feed gases at CO_2 partial pressures in excess of 0.5 MPa and system operating pressures of 1–10 MPa. Solubility and mass transfer can be optimized by providing a large interfacial contact area (i.e. tower packing), good mixing and sufficient contact time.

The solvent can be regenerated by a little heating or pressure reduction.

No chemical reactions occur, making the process stable and the energy requirement of regeneration lower than that for chemical absorption, as no chemical bonds need to be broken.

Water is a poor solvent for CO_2 and so a variety of liquid solvents, principally alcohols, ethers and ketones, are used, including polyethylene glycol as well as methanol under a range of trade names.

Hybrid physical/chemical absorption liquids are also in use.

6.4.3 Oxyfuel, Cryogenics and Chemical Looping

The collection of carbon dioxide from flue gases would be made more tractable if the presence of other gases could be reduced.

Atmospheric air is approximately 79% nitrogen and 21% oxygen by volume. The nitrogen serves no useful purpose in a combustion process and formation of its oxides produces further pollutants (NO_x).

Using pure oxygen in a *stoichiometric* hydrocarbon combustion process would produce carbon dioxide and water vapour only as products of combustion. The water vapour could be separated from the carbon dioxide by simple condensation. Carbon dioxide concentrations of 85–90% by volume in the resulting flue gas would then be possible.

This pre-combustion processing of air is the essence of the so-called *oxyfuel* process.

In practice, oxygen can be separated from air with a purity of 95–99%. The remaining few percent comprise some nitrogen and inert gases like argon.

Oxygen is produced on an industrial scale by *cryogenic air separation*. Air is compressed and cooled to about 93 K, at which point the oxygen condenses out. This is essentially a de-nitrogenation process.

A standard power-generation plant using the oxyfuel process will need modification. Due to the absence of nitrogen (which acts as a heat ballast), the use of pure oxygen raises the adiabatic flame temperature of the combustion process. Potential overheating is moderated by the recirculation of a fraction of the flue gases (see Figure 6.6).

Cooled flue gas recirculation to the burner may be needed to alleviate the problem in a retrofit situation. The burner may therefore be asked to operate in an oxygen/carbon dioxide/fuel environment. Optimization of burner design may be required, with excess oxidant implications.

However, with increased temperature come a lower heat exchange area, smaller size and cost savings.

Boiler plants lend themselves to retrofit more readily than gas turbine systems.

However, oxygen production comes at a price. Current estimates are that the oxyfuel system can consume up to 25% of a plant's electrical energy production, perhaps reducing plant efficiency by up to 10%. Much research is being carried out to produce an oxygen separation method with a reduced energy demand. Candidates include species-selective membranes and chemical sorbents.

Chemical looping combustion (CLC) is another pre-combustion process based on the ability of a metal (e.g. nickel, cobalt, copper, iron or manganese) to be readily oxidized

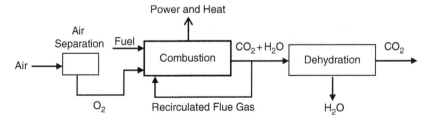

Figure 6.6 Recirculation of flue gases in an oxyfuel combustion system.

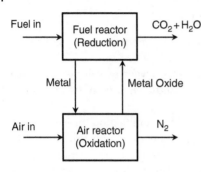

Figure 6.7 Chemical looping system.

in the presence of atmospheric oxygen and to release its oxygen, i.e. be reduced, in the presence of the fuel. A schematic is shown in Figure 6.7.

Denoting the metal M, in the air reactor:

$$M_xO_{y-1} + \frac{1}{2}O_2 \rightarrow M_xO_y$$

In the fuel reactor:

$$(2a + b)M_xO_y + C_aH_{2b} \rightarrow (2a + b)M_xO_{y-1} + aCO_2 + bH_2O$$

The metal oxygen carriers take the form of 0.1–0.5 mm particles. The reactor typically takes the form of a circulating fluidized bed, typically operating at a temperature of around 1000 °C.

The search is on for oxygen-carrying materials with good oxygen transfer capacities, high reactivity and rapid take-up. Other desirable characteristics include low fragmentation, low attrition and a low tendency to agglomerate. Low toxicity is also, of course, desirable.

Operational experience of CLC plant utilizing natural gas with a capacity of up to 50 kW was available around ten years ago. Some plant experience with solid fuel and syngas has become more available since 2010. So far, more than 900 suitable oxygen-carrying materials have been identified and about 30 (for example, NiO, Fe_2O_3, CuO, CoO) have been tested in practice.

6.4.4 Gas Membranes

Gas membrane technology is over 50 years old. Membranes separate gases by virtue of the presence of small pores or sorption and diffusion.

Carbon dioxide can be separated from a mixture of gases by exposing the mixture to a barrier or membrane that is selectively permeable to CO_2.

Current membrane technology is not as capture effective as absorption or cryogenics, and resulting CO_2 purities are inferior to these technologies. Membrane lifetimes are of the order of 3–5 years. However, energy requirements can be lower than for other technologies.

The driving force for separation is upstream (*feed*) to downstream (*permeate*) membrane partial pressure difference for the CO_2. Target gas remaining in the feed stream is often termed *retentate*.

6.4.4.1 Membrane Flux

As mentioned, the transport mechanisms through membranes are solution and diffusion or permeation through the empty space of the membrane.

Defining S_A as a *solubility coefficient* (m^3(STP)/m^3. m. Pa) of the gas (A) to be separated and D_A as the *diffusion coefficient* (m^3/s) or molecular mobility for the membrane, the *permeability* of the membrane to gas component A is given by the product:

$$\text{Permeability}_A = S_A D_A \tag{6.26}$$

The term *permeance* (m^2/s. Pa) or molecular conductance of the filter to gas (A) is also in use and is given by:

$$\text{Permeance}_A = \text{Permeability}_A / L \tag{6.27}$$

where L is the membrane thickness (m).

Clearly it is desirable to maximize this parameter by choosing a membrane material with a high permeability at the minimum membrane thickness possible commensurate with other operational conditions.

The volume flow rate per unit area of the membrane or *flux* ($\dot{V}_A$, m^3/s.m^2) of component gas (A) through the membrane is then given by:

$$\dot{V}_A = \text{Permeance}_A \times (P_{A,\text{feed}} - P_{A,\text{permeate}}) \tag{6.28}$$

where P_A is the partial pressure of the gas (A) to be separated.

A further membrane characteristic called the *selectivity* is in common use.

In a system with two gases this is defined as the ratio of the target gas permeability to the unwanted gas component permeability.

6.4.4.2 Maximizing Flux

Low permeance can be overcome by increased membrane area, and low values of selectivity can be overcome with multi-staging or cascading of membranes.

Looking at the flux relationship, the rate of transfer could also be increased by increasing the partial pressure of the target gas on the feed side of the membrane and/or lowering the partial pressure of the target on the permeate side (i) or by applying a vacuum (ii) or by *sweeping* the permeate side with an inert gas (iii) (see Figure 6.8).

6.4.4.3 Membrane Types

Membrane types can be divided into polymer and inorganic varieties.

Polymer (Polymeric) Membranes Polymer membranes used for pre- and post-combustion applications can be subdivided into polyimides, facilitated transport, mixed matrix and poly-ether-oxide (PEO) types.

Polyimides make good separation membranes because, in addition to a high selectivity and permeability for CO_2, they have a high thermal stability, good chemical resistance, mechanical strength and good electrical characteristics.

Facilitated transport membranes contain a metal ion carrier to chaperone a gas selectively across the barrier. Problems with carrier degradation leave a question mark over their use with coal-fired plant.

Mixed matrix membranes (MMMs) have a micro or nanoparticle inorganic material as part of a polymer matrix. At present, these types display only moderate separation

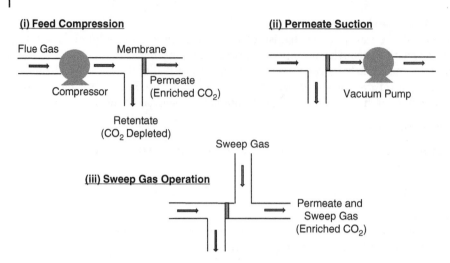

Figure 6.8 Membrane partial pressure enhancement strategies.

performance due to poor interfacial adhesion between matrix and inorganic component. Higher thicknesses are required, leading to a low flux.

Poly-ether-oxide (PEO) membranes have increased performance as a consequence of the polarity of the ether oxide presence. Good CO_2/N_2 selectivity and high CO_2 permeance have been reported.

Inorganic Membranes   Inorganic membranes use either (conductive) metals or ceramics as a selective barrier.

Metallic membranes tend to be described as *dense*. Palladium and its alloys are in use as membrane materials.

Ceramic membranes can be either *dense* or *microporous*. Dense ceramic membranes are used to facilitate either oxygen separation using ionic/electronic transfer (800–1000 °C) or hydrogen separation using protonic/electronic transfer (500–800 °C). Microporous membranes can be further subdivided into crystalline types (e.g. zeolites) and amorphous types. Structurally microporous membranes are graded. The smallest pore sizes need to be very small, as the diameters of gas molecules are typically less than 1 nm across and have a narrow range of values, for example, carbon dioxide 0.33 nm, hydrogen 0.289 nm, oxygen 0.346 nm and nitrogen 0.364 nm.

Operating temperatures for ceramic membranes are much lower (< 400 °C).

Both dense and microporous types are suitable for pre- (H_2/CO_2) and post- (CO_2/N_2) combustion separation.

6.5   Aspects of CO_2 Conditioning and Transport

The capture of carbon dioxide may take place at large power-generating plants that are remote from the final destination CO_2 storage site, and so a large network of CO_2 pipelines along with integrated truck/ship transportation may be required. Carbon dioxide pipeline transportation technology is not new, and there is already operating experience albeit for a different reason.

For example, the USA has a CO_2 network of several thousand kilometres that utilizes naturally occurring geological reservoirs of CO_2 for enhanced oil recovery (EOR) in oil fields. The CO_2 (often along with water) is pumped into depleting oil fields to maintain pressure at depth to keep oil production at an acceptable level. Pipe diameters are typically less than 1 m.

Dedicated CO_2 pipelines performing the same function can also be found in Algeria, Canada, Hungary and Turkey.

However, the issues associated with this provision in the carbon sequestration chain are many.

6.5.1 Multi-stage Compression

Carbon dioxide is usually transported in a liquid or supercritical condition.

The advantages of the supercritical condition are a high density (smaller pipes) and a low viscosity (lower pressure drop/km).

In all cases, the occurrence of two-phase (gas/liquid) flow is to be avoided, as this can result in increased energy pumping costs and possible pipeline damage due to cavitation. For this reason, a minimum pressure above critical conditions (73.9 bar) is maintained in existing pipelines.

Minimizing the work input or energy associated with CO_2 pressurization is important. The work input to a compression process will be minimized when irreversibilities, for example, heat losses, friction, noise etc., are minimized. The (internally) reversible work input will then be given by:

$$w_{rev,in} = \int_1^2 v dP \tag{6.29}$$

The minimum work could be achieved by employing an excruciatingly slow compression, but it is more practical to minimize the specific volume by keeping the gas temperature as low as possible during compression. The ideal case would then be an *isothermal compression*.

A comparison of an *isothermal compression process* (max cooling, index of compression 1) with an *isentropic compression process* (no cooling, index of compression k) and a *polytropic compression process* (some cooling, index of compression, $1 < n < k$) assuming perfect gas conditions is shown in Figure 6.9.

Remember that the work of compression is the area under the curve on a P–v diagram. The work (kJ/kg) expressions associated with each process are:

Isentropic:

$$w_{comp,in} = \frac{kR(T_2 - T_1)}{k - 1} = \frac{kRT_1}{k - 1}\left[\left(\frac{P_2}{P_1}\right)^{(k-1)/k} - 1\right] \tag{6.30}$$

Polytropic:

$$w_{comp,in} = \frac{nR(T_2 - T_1)}{n - 1} = \frac{nRT_1}{n - 1}\left[\left(\frac{P_2}{P_1}\right)^{(n-1)/n} - 1\right] \tag{6.31}$$

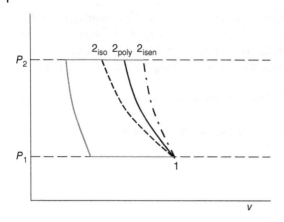

Figure 6.9 Comparison of gas compression (1–2) processes.

Isothermal:

$$w_{comp,in} = RT \ln \frac{P_2}{P_1} \tag{6.32}$$

For a single-stage compression, a cooling jacket around the compression cylinder may be sufficient, but for multi-stage systems, constant pressure intercooling between stages is employed, with the fluid being cooled to the compressor inlet temperature.

Although real compression processes are usually regarded as polytropic, intercooling and staging do reduce the work input, as shown in Figure 6.10.

The total work for a two-stage polytropic compression process (index of compression n) is given by:

$$w_{comp,in} = \frac{nRT_1}{n-1} \left[\left(\frac{P_x}{P_1} \right)^{(n-1)/n} - 1 \right] + \frac{nRT_2}{n-1} \left[\left(\frac{P_2}{P_x} \right)^{(n-1)/n} - 1 \right] \tag{6.33}$$

where P_x is the intermediate pressure and T_2 is the temperature of the gas entering the second stage. Note that with perfect intercooling between stages, $T_1 = T_2$.

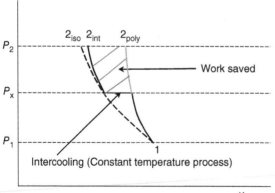

Figure 6.10 Gas compression with intercooling.

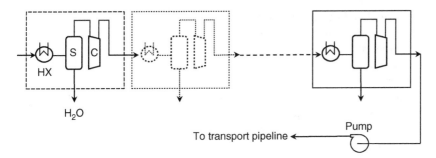

Figure 6.11 Dehydration/pressurization cascade.

The work saved is a function of the intermediate pressure. The optimal intermediate pressure that minimizes the work input and maximizes the work saved can be determined by differentiating the above expression with respect to P_x and setting the result to zero. This gives:

$$P_x = (P_1 P_2)^{\frac{1}{2}} \text{ or } \frac{P_x}{P_1} - \frac{P_2}{P_x} \qquad (6.34)$$

Typically, the CO_2 is pressurized in a multi-stage process to just above critical pressure (~80 bar) and pumped up to the transport pressure (~150 bar). A typical dehydration/pressurization string comprising a heat exchanger (HX) or cooler, water separator (S) and compressor (C) is shown in Figure 6.11.

6.5.2 Pipework Design

6.5.2.1 Pressure Drop
A significant drop in pressure will occur over long pipe lengths. It is undesirable for the pressure changes to result in two-phase conditions and so pumping stations will be required along the length of the line.

Pressure losses for single-phase flow can be approximated using a range of standard techniques but will generally depend on pipe diameter and material, CO_2 density and viscosity and flow rates and velocities. Because of the high-pressure operating environment, increased pipe thicknesses are specified.

6.5.2.2 Materials
Carbon dioxide forms carbonic acid with water, but if the CO_2 is dry then carbon steel pipes are suitable. Large temperature reductions can occur under depressurization and so pipes must also be fracture-resistant.

Seal/gasket material selection should be given special consideration, as standard elastomers absorb CO_2 and swell, resulting in potential failure.

Plastics such as PTFE, PP and nylon are generally suitable for use with CO_2.

Hydrocarbon- and fluorocarbon-based lubricants are used with CO_2 systems.

6.5.2.3 Maintenance and Control
Pipelines are monitored at intervals and controlled centrally with respect to pumping speeds, valve operation, leak detection, flow rate measurement etc.

Flow measurement and metering techniques include orifice and vortex types.

The internal pipe conditions can be monitored with the use of *pigs* performing a cleaning function as well as checking for internal corrosion and mechanical deformation.

Pipelines should also be designed with sufficient isolation, emergency shut-off and blow down valves.

For smaller quantities of CO_2, ship transport is also an option. In this case, the CO_2 is usually transported as a chilled liquid.

Current CO_2-dedicated ship transports are few and small in capacity (typically 1000–1500 m^3). There are a few larger vessels with a capacity of up to 10 000 m^3 certified to carry CO_2. Transport pressures of up to 20 bar are used.

6.5.3 Carbon Dioxide Hazards

6.5.3.1 Respiration

Inhaled (atmospheric) carbon dioxide has a concentration of about 0.04%. The partial pressure of carbon dioxide in the blood is higher than this (1.4%) and so CO_2 can diffuse out of the blood to be exhaled. If the concentration in inhaled air increases above this level, carbon dioxide cannot be released from the blood. In fact, further CO_2 may be absorbed, increasing blood acidity. The involuntary response may be to take longer/deeper breaths, exacerbating the situation.

The *time-dependent* physiological effects of increasing ambient CO_2 concentration (%) are divided into four categories of increasing severity. For example, at an exposure time of around 20 minutes, indicative effects at associated ambient concentrations (%) are as follows:

0–1% – no noticeable effect
1–4.5% – small hearing loss, doubling in depth of respiration
4.5–7.5% – mental depression, dizziness, nausea
>7.5% – dizziness, stupor, unconsciousness.
Exposure limits in the UK are set by the Health and Safety Executive. Current values (see www.hse.gov.uk/carboncapture/carbondioxide.htm) are:Long term (8 hr) – 5000 ppm, i.e. 0.5% CO_2
Short term (15 min) – 15 000 ppm, i.e. 1.5% CO_2.

6.5.3.2 Temperature

If gaseous carbon dioxide experiences a sudden drop in pressure, cooling will result. Carbon dioxide snow may be formed, causing cold burns when coming into contact with skin or eyes.

6.5.3.3 Ventilation

In the event of a leak, high concentrations of CO_2 can build up quickly in confined spaces. For example, the specific volume of liquid CO_2 will increase by approximately 800 times on evaporation.

The molar mass of CO_2 is greater than air, hence carbon dioxide will tend to accumulate at low levels in an unventilated, still air situation. Detection equipment and low-level exhaust louvres are necessary in such spaces and around plant.

Due to the reduction in temperature associated with a pressure drop, contaminants such as water may also condense, forming fogs. Although useful in indicating a leak, they are no indication of resulting CO_2 concentration.

6.6 Aspects of CO_2 Storage

Storage of carbon dioxide in pressure vessels is out of the question because of the magnitude of annual CO_2 combustion-associated release. However, Nature has been practising carbon capture for quite some time, and enhancing these techniques provides at least a direction for solving the CO_2 emission problem.

6.6.1 Biological Sequestration

In 1660, Flemish chemist Jan Baptista van Helmont started to grow a willow tree in a pot. After five years the tree weighed about 70 kg while the soil had only lost about 30 g in mass. He concluded that the soil matter (mostly aluminium silicates) was largely irrelevant to the growth of plants other than to provide mechanical support and act as a water-storage vessel.

The soil water, of course, is vital in itself and as a medium for the supply of very small amounts of dissolved inorganic substances needed by plants.

Even the supply of water does not provide the whole story. Living matter contains much carbon but there is insufficient carbon in soil to account for plant growth.

As fate would have it, van Helmont had also conducted mass balance experiments on the burning of charcoal. He noted that the post-combustion ash weighed less than the fuel, and came to the conclusion that some *wild spirit* or *gas* had been formed to account for the mass difference.

In 1727, English physiologist Stephen Hales determined that atmospheric carbon dioxide was the source growth material that van Helmont had been looking for.

In the 1770s, English minister Joseph Priestly carried out some simple experiments with a bell jar. First, he attempted to grow a plant under a bell jar with a limited supply of carbon dioxide. He noted that even if water and sunlight were plentiful, the plant would grow no more once its supply of CO_2 was exhausted. Next, he placed a mouse under a bell jar and noted that it would expire when its supply of oxygen (O_2) had been exhausted. Finally, he placed a mouse and a plant under a bell jar together and noted that they would continue to survive for longer than each did in isolation. From these results, he correctly reasoned that plants on land (and in the oceans) consumed carbon dioxide and released oxygen while animals consumed oxygen and gave off carbon dioxide.

The significance of this was not lost on Dutch physician Jan Ingenhousz, who, in 1779, published one of the earliest books on ecology and material balances in the environment. His ideas about carbon dioxide, oxygen and water utilization and reformation became crystallized by the phrase *photosynthesis*, meaning 'manufactured by light'.

Photosynthesis is a widely used term to describe how green plants use carbon dioxide, water and sunlight to manufacture carbohydrates (or *starch*). A simple equation describing this process is given below:

$$CO_2 + H_2O \rightarrow (CH_2O) + O_2$$

In the 19th century, work was done on determining which characteristics of plants facilitated this process. Attention was focused on what made plants green, as this colouration is uncommon in animals. In 1819, two French chemists, Pierre Joseph Pelletier and Joseph Bienaime Caventou, succeeded in extracting a green pigment called *chlorophyll* (meaning 'green leaf'). They determined that chlorophyll is, in fact, a *catalyst*.

The structure of chlorophyll was not established until the early part of the 20th century, when two German chemists, Richard Wilstatter and Hans Fischer, determined its composition.

Chlorophyll has a complex structure with a number of variants, but essentially it comprises a ring-shaped *head* with a *magnesium atom* (Mg) at its centre bonded to four nitrogen atoms. The head is connected to a long hydrocarbon molecule 'tail', known as a *carotenoid* (see Figure 6.12).

The substance appears green because it absorbs the red, orange and blue wavelengths (450–670 nm) of light whilst reflecting the remainder. These longer wavelengths contain less energy than light at the blue end of the spectrum but are able more easily to penetrate the dust and gas of the atmosphere. The solar energy absorbed by the chlorophyll raises its electrons from their 'ground' state to an elevated level. On return to the ground state, the energy is made available to facilitate chemical reactions.

These reactions were first discovered by experiments with isotopic tagging using heavy oxygen (O^{18}). The work was carried out by two American biochemists, Samuel Ruben and Martin D. Kamen. They determined that it was the water molecule that was split up into hydrogen and oxygen during photosynthesis. Photosynthesis is therefore concerned with the *photolysis*, or breaking down, of water.

The split oxygen and hydrogen components of the water can have two possible futures. Half of the hydrogen takes part in the so-called *respiratory chain*, generating three molecules of a chemical energy store called adenosine triphosphate (ATP) to be reunited later with some of the oxygen to form water. The remainder of the hydrogen combines with the carbon dioxide to form carbohydrates. This is facilitated by the presence of a five-carbon sugar molecule, ribulose bisphosphate (RuBP), found in plant cells. This process is energy consuming and utilizes the previously manufactured ATP.

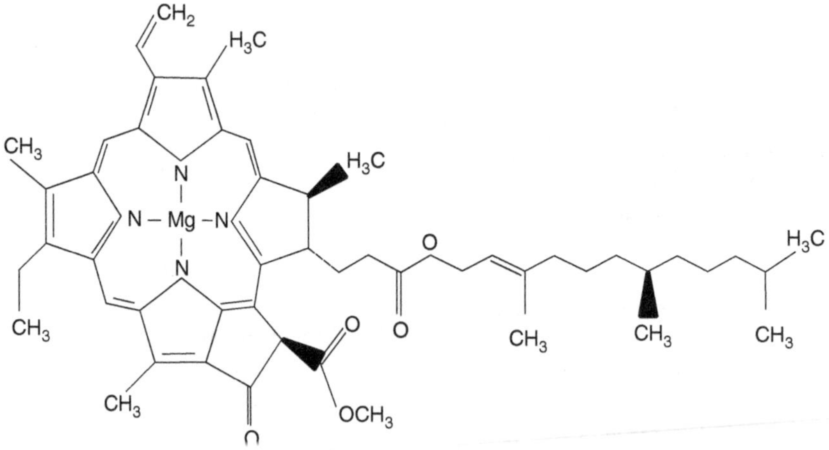

Figure 6.12 Structure of chlorophyll.

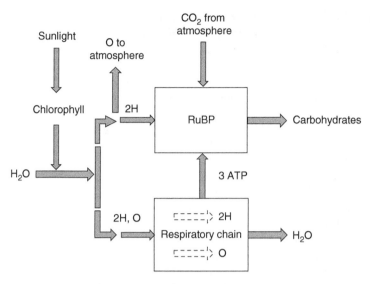

Figure 6.13 Simplified schematic of photosynthetic pathway.

Oxygen is not used in this latter process and is released to the atmosphere. This is summarized in Figure 6.13.

In this way, land-based and oceanic plants have been extracting atmospheric CO_2 for millions of years. Clearly, not destroying existing planetary forestation and vegetation and enhancing and encouraging replanting will help. However, it is widely believed that this approach cannot alone supply a carbon sink sufficient to compensate for anthropogenic combustion activity.

6.6.2 Mineral Carbonation

Carbon dioxide has an affinity to combine with some light metal oxides like magnesium and calcium to form solid metal carbonates, for example:

$$MgO + CO_2 \rightarrow MgCO_3$$

The reaction is exothermic, i.e. it releases heat rather than requiring it.

Simple light metal oxides are rare in the lithospheric environment and these metals are more commonly found in silicate combinations such as in the minerals forsterite (Mg_2SiO_4) and serpentine ($Mg_3Si_2O_5(OH)_4$).

These, too, however, will combine with carbon dioxide, thus:

$$\frac{1}{2}Mg_2SiO_4 + CO_2 \rightarrow MgCO_3 + \frac{1}{2}SiO_2$$

$$\frac{1}{3}Mg_3Si_2O_5(OH)_4 + CO_2 \rightarrow MgCO_3 + \frac{2}{3}SiO_2 + \frac{2}{3}H_2O$$

The resulting carbonate products are stable and do not re-release the captured CO_2. This is in contrast to other carbon storage options where leakage is a possibility.

These reactions have, therefore, been proposed as the basis of a carbon capture and storage solution, since the raw material magnesium-rich rocks are relatively common and close to the surface, making their extraction relatively simple.

However, resulting from the prevailing reaction kinetics, simply exposing carbon dioxide to rocks directly would be an unacceptably slow process and, in practice, the rock would need pre-treatment in the form of crushing and the addition of a solvent to precipitate out the desired minerals.

The solvation itself presents problems, as current solvent candidates include hydrochloric acid, molten magnesium chloride salts and water. Using an acid solvent is an energy-hungry process, requiring the removal of large quantities of water. The molten salt option brings corrosive problems to the process. Using water (enhanced with Na_2CO_3) as a solvent is, on the face of it, less environmentally aggressive but is slow and still energy-intensive.

To date, therefore, all mineral carbonation processes would impose a significant parasitic effect on the energy produced by the target power station.

Due to low overburden and the presence of comparatively thick mineral-bearing strata, the mining cost of metal silicates appears to be low. It has been estimated that typically around 8 tonnes of magnesium silicate ore would be needed to capture and store the carbon dioxide released by 1 tonne of coal.

The infrastructure needed to transport such large quantities of raw material to a distant power station would be significant, and it has been proposed that, if this technology can be made economic, the capture process be carried out at the mineral source mine with storage on site. For minimum impact, the CO_2 generator would, of course, then be sited adjacent to the mineral mine.

6.6.3 Geological Storage Media

Geological or fossil CO_2 is found in rocks at a number of sites around the world, for example, Mexico, the USA, Hungary, Turkey and Romania.

CO_2 is also very commonly associated with oil and natural gas reservoirs. Indeed, there are so-called low-BTU gas fields in the USA and Asia comprising mostly CO_2 in association with some methane and low levels of other gases like H_2, SO_x, N_2 and C_3H_8.

This occurrence would indicate that one way forward for large-scale carbon dioxide storage is to inject the gas into suitable rock strata.

At the time of writing, attention is focused on porous media such as depleted oil and gas reservoirs and deep saline rocks.

The important characteristics for a potential porous store are:

- Local and regional geology – long-term tectonic stability.
- Presence of an impermeable, continuous overlying or 'cap' rock.
- Storage dimensions and porosity – dependent on overall vertical and horizontal dimensions.
- Storage pressure – flow rate into (injectivity) and through (permeability) the store.

Carbon dioxide is best injected in its supercritical condition to provide a high density associated with a low viscosity.

The supercritical pressure (>7.3 MPa) can be maintained at depths of about 800 m.

Geological trapping can be subdivided into structural, stratigraphic, solubility, residual and adsorption forms.

- *Structural trapping* relies on the presence of an impermeable cap rock to halt the upward migration of the CO_2.

This can be provided by the presence of *anticlinal* (arch-shaped) or *faulted* (slipped) rock formations (see Figure 6.14).

- *Stratigraphic trapping* relies on changes in the porosity and permeability characteristics within a given rock stratum.
- *Solubility (or dissolution) trapping* relies on the CO_2 dissolving into the existing rock fluids. The solubility depends on pressure, temperature, contact area and composition of the inherent solvent.
- *Residual trapping* is based on the *drainage-imbibition* process. In this phenomenon, the carbon dioxide will displace the existing strata fluids (brine or hydrocarbons) but only if the injection pressure is maintained. After this ceases, the displaced fluid attempts to return to the rock, thus encasing the CO_2 and forming a seal around the gas. The process depends on the rock pore capillary forces.
- *Adsorption trapping* of CO_2 storage in coal seams relies on the adsorption of the gas, displacing the *in situ* methane from pores within the rock. The process is reliant on concentration-dependent diffusion.

  It has been determined that the suitability of a coal seam for storage is dependent on a number of factors including:

  - ➢ Coal carbon and volatile content.
  - ➢ Temperature: Adsorbed gas decreases with increasing temperature.
  - ➢ Water content: Adsorbed gas decreases with decreasing water content.
  - ➢ Ash content: Adsorbed gas decreases with increasing water content.
  - ➢ Desorbed gas composition.
- *Salt domes* (see Figure 6.15) provide another option for geological gas storage. Salt domes are essentially column-shaped aquifers (containing non-potable water or brine in their pores) that have deformed and intruded into overlying layers. They generally comprise sodium chloride with some calcium sulphate.

Pumping water into the formation has a leaching effect and creates cavities suitable for gas storage.

The technology is not new and was deployed first (France, Germany, South Africa and the USA) in the 1970s to store natural gas and oil during the energy crises.

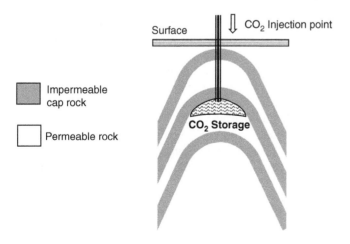

Figure 6.14 Example of structural trapping in an anticline.

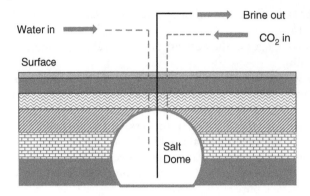

Figure 6.15 Salt dome CO_2 storage.

If the walls of the dome are impervious, the system can be leak-tight.

A crude estimate of the CO_2 storage capacity (m_{CO_2}, kg) of a geological feature is given by:

$$m_{CO_2} = A_{strata} \times h_{strata} \times \text{rock porosity} \times \eta_{storage} \times \rho_{CO_2} \tag{6.35}$$

where $\eta_{storage}$ is a measure of the injection effectiveness. Typical efficiencies range from 2–40%.

6.6.4 Oceanic Storage

The oceans cover an area of approximately 362×10^6 km^2, i.e. 71% of the earth's surface, and provide a habitat for about 80% of the planet's life.

The average depth of the oceans is about 4 km. This compares with an average height above sea level of less than 1 km for the land masses. At the deepest point, in the Marianas Trench in the Pacific Ocean, the land falls away to more than 11 km.

Typically, the breakdown by volume for dissolved gases in the ocean indicates a reduced nitrogen content and an increase in oxygen and carbon dioxide presences compared to atmospheric air. Reported values vary.

Although surface waters are often saturated with atmospheric gases, Henry's Law is of limited use in modelling actual hydrospheric conditions in deep water. Here, conditions are usually determined by measurement.

The distribution of dissolved gases in the ocean is uneven, having a layered structure linked to winds, waves, oceanic currents and biological activity such as photosynthesis, respiration and decomposition.

Levels of carbon dioxide are at a minimum of about 60–70 mg/litre at the surface, due to consumption by marine plants in photosynthesis, and increase to a maximum value at around 1000 m depth due to respiration and decay generally.

Carbon dioxide dissolved in water generates ions. The addition of carbon dioxide to oceanic water therefore increases its acidity:

$$CO_2(aq) + H_2O \rightarrow H_2CO_3(aq)$$

$$H_2CO_3(aq) \rightarrow H^+ + HCO_3^-$$

$$HCO_3^- \rightarrow H^+ + CO_3^{2-}$$

As a consequence of anthropogenic activity and related atmospheric carbon dioxide concentrations, it is estimated that the planet's oceanic surface water pH value has decreased (i.e. acidity has increased) by 0.1 to under 8.1 since the beginning of the Industrial Revolution. Notwithstanding this, the ocean has been suggested as a potential store for the direct injection of CO_2 emissions.

Predicting the state of carbon dioxide is dependent on knowledge of its pressure and temperature. The pressure and temperature conditions in the ocean, i.e. the CO_2 containment vessel, vary markedly.

Water pressures in the oceanic hydrosphere vary from zero (gauge) at a free surface to over 100 MPa at great oceanic depth, i.e. approximately 1000 × that found at sea level (~1 × 10^5 Pa). The exact pressure–depth relationship in deep water is, however, nonlinear due to changes in composition and, at great depth, some slight compressibility effects.

The oceanic hydrosphere is, in some instances, heated by a range of local, high-intensity temperature sources such as volcanoes, hot springs etc. For example, local water temperatures of up to 400 °C have been found at hydrothermal vents on the sea floor.

However, the main source of energy input, especially for open bodies of water, is solar radiation. Solar radiation at the earth's surface is a diffuse form of energy. Average oceanic *surface* temperatures are typically 18 °C, with a range of −2 to 36 °C.

Over its entire volume, however, the ocean is a relatively cold place, with its average temperature in the range of 3–4 °C. Global ocean temperatures vary with latitude and depth, largely due to differences in solar energy input and wave/current action. The coldest water is at the poles; the warmest between the tropics.

With this in mind, CO_2 oceanic injection points at depths of less than 500 m result in the CO_2 existing in its gaseous form.

At depths of between 500 m and around 2700 m, the CO_2 will exist as a liquid but will be positively buoyant, i.e. it will have a tendency to float up, being less dense than the surrounding seawater.

Below 2700 m, the CO_2 will be more dense than the surrounding seawater and will be negatively buoyant, i.e. it will tend to sink.

In addition to gaseous and liquid carbon dioxide, a third state is of interest – the solid CO_2 hydrate. Carbon dioxide hydrates are created when the hydrogen bonds of adjacent water molecules form a framework with a central cavity occupied by CO_2 molecules. The attraction of this exotic structure is its stability, negative buoyancy and its ability to promote dispersion. It is, however, at present difficult to produce, requiring relatively low temperatures (~10 °C) and high pressures.

Current suggested methods of CO_2 release into the ocean include forming either CO_2 lakes on the oceanic floor (>3000 m) or CO_2 plumes within the oceanic body itself. Both shore-sourced pipelines and ship transport/offshore delivery systems are being considered.

Whatever the chosen method, the CO_2 is not permanently isolated from the atmosphere and, over time (several centuries), the gas will be liberated and reunited with the atmosphere via the ocean surface. However, it is hoped that man-made release of carbon dioxide will have abated by that time.

In terms of environmental impact, the effects on marine organisms of the increase in CO_2 oceanic water concentrations and attendant lowered pH values associated with direct injection are under intensive study.

6.7 Worked Examples

Data for use in the following worked examples:

Universal gas constant: $R_o = 8.314$ kJ/kmol K
Molar mass of carbon dioxide: $M_{CO_2} = 44$ kg/kmol
Molar mass of nitrogen: $M_{N_2} = 28$ kg/kmol
Carbon dioxide (CO_2) critical point properties: 73.9×10^5 Pa, 304.2 K
Nitrogen (N_2) critical point properties: 3.39 MPa, 126.2 K.

Worked Example 6.1 – Use of compressibility chart
Using a compressibility chart, determine the specific volume (m^3/kg) of carbon dioxide at critical pressure and 349.83 K.
   Take the specific gas constant for carbon dioxide to be 188.9 J/kg K.

Solution
Given: $P_{cr,CO_2} = 73.9 \times 10^5$ Pa, $T = 349.83$ K, $T_{cr,CO_2} = 304.2$ K, $R = 188.9$ J/kg K.

Find v.

$$P_r = 1, \quad T_r = 349.83/304.2 = 1.15$$

From the compressibility chart, $z = 0.75$.

$$Pv = zRT$$

$$73.9 \times 10^5 \times v = 0.75 \times 188.9 \times 349.83$$

$$v = 6.71 \times 10^{-3} \text{ m}^3/\text{kg}.$$

Worked Example 6.2 – Mass and molar properties of a gas mixture
A gas mixture comprises 88 kg of carbon dioxide (CO_2) and 364 kg of nitrogen (N_2).
   Determine the mass fraction and molar fraction for each gas and the average molar mass (kg/kmol) and gas constant (kJ/kg K) for the mixture.

Solution
Given: m_{CO_2}, m_{N_2}, M_{CO_2}, M_{N_2}, R_o.

Find mf_{CO_2}, mf_{N_2}, y_{CO_2}, y_{N_2}, M_{mix}, R_{mix}.

Number of moles of carbon dioxide $= m_{CO_2}/M_{CO_2}$
$$= 88/44 = 2 \text{ kmol}$$

Number of moles of nitrogen $= m_{N_2}/M_{N_2}$
$$= 364/28 = 13 \text{ kmol}$$

Mass fractions

$$m_{mix} = m_{CO_2} + m_{N_2} = 88 + 364 = 452 \text{ kg}$$

$$mf_{N_2} = 364/452 = 0.805 \qquad mf_{CO_2} = 1 - mf_{N_2} = 0.195$$

Molar fractions

$$n_{mix} = 2 + 13 = 15 \text{ kmol}$$

$$y_{N_2} = 13/15 = 0.87 \qquad y_{CO_2} = 1 - y_{N_2} = 0.13$$

Average molar mass:

$$M_{mix} = m_{mix}/n_{mix} = 452/15 = 30.1 \text{ kg/kmol}$$

Average gas constant:

$$R_{mix} = R_o/M_{mix} = 8.314/30.1 = 0.276 \text{ kJ/kg K}$$

Worked Example 6.3 – Determination of gas mixture partial pressures

A sample of flue gas at a pressure of 1 bar has the following composition:

1 kmol of CO_2, 2 kmol of H_2O, 0.317 kmol of O_2 and 8.713 kmol of N_2.

Determine the partial pressure (Pa) of each component gas.

Solution

Given: $n_{CO_2} = 1$ kmol, $n_{H_2O} = 2$ kmol, $n_{O_2} = 0.317$ kmol, $n_{N_2} = 8.713$ kmol, $P_{total} = 1$ bar $= 1 \times 10^5$ Pa.

Find P_{i, H_2O}, P_{i, O_2}, P_{i, N_2}.

$$n_{total} = n_{CO_2} + n_{H_2O} + n_{O_2} + n_{N_2}$$

$$n_{total} = 1 + 2 + 0.317 + 8.713 = 12.03 \text{ kmol}$$

$$P_i = P_{total} \frac{n_i}{n_{total}}$$

$$P_{i,CO_2} = 1 \times 10^5 (1/12.03) = 8313 \text{ Pa}$$

$$P_{i,H_2O} = 1 \times 10^5 (2/12.03) = 16\,625 \text{ Pa}$$

$$P_{i,O_2} = 1 \times 10^5 (0.317/12.03) = 2635 \text{ Pa}$$

$$P_{i,N_2} = 1 \times 10^5 (8.713/12.03) = 72\,427 \text{ Pa}.$$

Worked Example 6.4 – Comparison of gas modelling laws (I)

A fixed-volume container is to hold a gas mixture comprising 2 kmol of carbon dioxide (CO_2) and 13 kmol of nitrogen (N_2) at a mixture pressure of 10 MPa and temperature of 300 K.

Determine the container volume using:

The ideal gas law.
Kay's rule/compressibility chart.

Solution

Given: n_{CO_2}, n_{N_2}, M_{N_2}, M_{CO_2}, P_{mix}, T_{mix}.

Find V_{mix} using the ideal gas law and Kay's rule.

Molar fractions

$$n_{mix} = 2 + 13 = 15 \text{ kmol}$$

$$y_{N_2} = 13/15 = 0.87, \qquad y_{CO_2} = 2/15 = 0.13$$

Ideal gas law

$$P_{mix} V_{mix} = n_{mix} R_o T_{mix}$$

$$10 \times 10^6 \times V_{mix} = 15 \times 8.314 \times 10^3 \times 300$$

$$V_{mix} = 3.714 \ m^3$$

Kay's rule

Calculating pseudocritical values:

$$P_{cr}^* = \sum (y_i P_{cr}) = (y_{N_2} \times P_{cr,N_2}) + (y_{CO_2} \times P_{cr, CO_2})$$

$$P_{cr}^* = (0.87 \times 3.39) + (0.13 \times 7.39) = 3.91 \text{ MPa}$$

$$T_{cr}^* = \sum (y_i T_{cr}) = (y_{N_2} \times T_{cr,N_2}) + (y_{CO_2} \times T_{cr, CO_2})$$

$$T_{cr}^* = (0.87 \times 126.2) + (0.13 \times 304.2) = 149.34 \text{ K}$$

$$T_r = T_{mix}/T_{cr}^* = 300/149.34 = 2.0$$

$$P_r = P_{mix}/P_{cr}^* = 10/3.91 = 2.56$$

From the compressibility chart, $z = 0.97$.

$$P_{mix} V_{mix} = z_{mix} n_{mix} R_o T_{mix}$$

$$10 \times 10^6 \times V_{mix} = 0.97 \times 15 \times 8.314 \times 10^3 \times 300$$

$$V_{mix} = 3.63 \text{ m}^3.$$

Worked Example 6.5 – Comparison of gas modelling laws (II)

A fixed-volume container is to hold a gas mixture comprising 2 kmol of carbon dioxide (CO_2) and 13 kmol of nitrogen (N_2) at a pressure of 10 MPa and temperature of 300 K.

Determine the container volume (m³) using Amagat's Law and compressibility factors.

Solution

Given: n_{CO_2}, n_{N_2}, M_{N_2}, M_{CO_2}, P_{mix}, T_{mix}.

Find V_{mix} using compressibility factors and Amagat's Law.

Molar fractions

$$n_{mix} = 2 + 13 = 15 \text{ kmol}$$

$$y_{N_2} = 13/15 = 0.87, \quad y_{CO_2} = 0.13$$

Amagat's Law

$$T_{r,N_2} = T_{mix}/T^*_{cr, N_2} = 300/126.2 = 2.38$$

$$P_{r,N_2} = P_{mix}/P^*_{cr, N_2} = 10/3.39 = 2.95$$

From the compressibility chart, $z_{N2} = 1.0$.

$$T_{r,CO_2} = T_{mix}/T^*_{cr, CO_2} = 300/304.2 = 0.99$$

$$P_{r,CO_2} = P_{mix}/P^*_{cr, CO_2} = 10/7.39 = 1.35$$

From the compressibility chart, $z_{CO_2} = 0.225$.

$$z_{mix} = (y_{N_2} \times z_{N_2}) + (y_{CO_2} \times z_{CO_2})$$
$$= (0.87 \times 1.0) + (0.13 \times 0.225)$$
$$= 0.9$$

$$P_{mix} V_{mix} = z_{mix} n_{mix} R_0 T_{mix}$$

$$10 \times 10^6 \times V_{mix} = 0.9 \times 15 \times 8.314 \times 10^3 \times 300$$

$$V_{mix} = 3.37 \text{ m}^3.$$

Worked Example 6.6 – Use of NFEE to determine mixture properties

A divider in a closed, adiabatic two-compartment vessel separates 8 kg of carbon dioxide at 200 kPa, 300 K and 2 kg of nitrogen at 100 kPa, 350 K.

If the divider is removed so that the gases can mix, determine the final temperature (°C) and pressure (kPa) in the unified vessel.

Assume ideal gas conditions apply.

Take the specific heat capacities of carbon dioxide and nitrogen at constant volume to be 0.657 kJ/kg K and 0.744 kJ/kg K, respectively.

Solution

Given: $m_{CO_2} = 8$ kg, $m_{N_2} = 2$ kg, $P_{CO_2} = 200$ kPa, $P_{N_2} = 100$ kPa, $T_{CO_2} = 300$ K, $T_{N_2} = 350$ K, $M_{CO_2} = 44$ kg/kmol, $M_{N_2} = 28$ kg/kmol, $C_{v, CO_2} = 0.657$ kJ/kg K, $C_{v, N_2} = 0.744$ kJ/kg K.

Find T_{mix}, P_{mix}.

From the first law of thermodynamics considerations, $W = 0$, $Q = 0$ (adiabatic), therefore $\Delta U = 0$.

That is

$$\Delta U_{CO_2} + \Delta U_{N_2} = 0$$

$$mC_v(T_{mix} - T_i)_{CO_2} + mC_v(T_{mix} - T_i)_{N_2} = 0$$

$$\{8 \times 0.657 \times (T_{mix} - 300)\} + \{2 \times 0.744 \times (T_{mix} - 350)\}$$

$$T_{mix} = 311 \text{ K} = 38\,°C.$$

Molar analysis

$$n_{CO_2} = m_{CO_2}/M_{CO_2} = 8/44 = 0.182 \text{ kmol}$$

$$n_{N_2} = m_{N_2}/M_{N_2} = 2/28 = 0.0714 \text{ kmol}$$

$$n_{mix} = n_{CO_2} + n_{N_2} = 0.2534 \text{ kmol}$$

Using the ideal gas law to calculate component volumes:
For the carbon dioxide (CO_2):

$$PV = nR_oT$$

$$200 \times 10^3 \times V_{CO_2} = 0.181 \times 8314 \times 300$$

$$V_{CO_2} = 2.26 \text{ m}^3$$

For the nitrogen (N_2):

$$PV = nR_oT$$

$$100 \times 10^3 \times V_{N_2} = 0.0714 \times 8314 \times 350$$

$$V_{N_2} = 2.08 \text{ m}^3$$

$$\text{and } V_{mix} = V_{CO_2} + V_{N_2} = 4.34 \text{ m}^3$$

Then:

$$P_{mix}V_{mix} = n_{mix}R_oT_{mix}$$

$$P_{mix} \times 4.34 = 0.2524 \times 8314 \times 311$$

$$P_{mix} = 150\ 968 \text{ Pa} = 150.97 \text{ kPa}.$$

Worked Example 6.7 – Entropy of gas mixture

3 kmol of carbon dioxide and 7 kmol of nitrogen are separated by a divider in a closed, adiabatic two-compartment vessel.

If the divider is removed and the gases are allowed to mix, determine the change in entropy (kJ/K).

Solution

Given: $n_{CO_2} = 3$ kmol, $n_{N_2} = 7$ kmol, $M_{CO_2} = 44$ kg/kmol, $M_{N_2} = 28$ kg/kmol.

Find ΔS_{mix}.

Molar analysis

$$n_{mix} = n_{CO_2} + n_{N_2} = 3 + 7 = 10 \text{ kmol}$$

Carbon dioxide mole fraction: $y_{CO_2} = 3/10 = 0.3$
Nitrogen mole fraction: $y_{N_2} = 7/10 = 0.7$
Then:

$$\Delta S_{mix} = -R_o \sum n_i \ln y_{i,2}$$

$$= -8.314 \times [(3 \times \ln(0.3)) + (7 \times \ln(0.7))]$$

$$= -8.314 \times [-3.611 + (-)2.497]$$

$$= 50.78 \text{ kJ/K}.$$

Worked Example 6.8 – Determination of work of separation

Determine the reversible work (kJ/kg_{CO_2}) associated with the removal of carbon dioxide having a partial pressure of 130 mbar in a gas mixture at 1 bar and a temperature of 60 °C.
Take the specific gas constant for carbon dioxide to be 188.9 J/kg K.

Solution

Given: $T_o = 60\ °C = 333\ K$, $R_{CO_2} = 188.9 J/kg\ K$, $P_i = 130 \times 10^{-3}$ bar, $P_{tot} = 1$ bar.

Find $W_{min,\ sep}$.

$$y_{CO_2} = P_{CO_2}/P_{tot}$$

$$= 130 \times 10^{-3}/1$$

$$= 0.13$$

$$W_{min,sep} = R_{CO_2} T_o \ln(1/y_{CO_2})$$

$$= 188.9 \times 333 \times \ln(1/0.13)$$

$$= 128.34 \text{ kJ/kg of } CO_2.$$

Worked Example 6.9 – Resourcing power plant CO_2 mineral carbonation

A large coal-fired power station emits 20 Mtonnes of CO_2 per annum. It is proposed to use forsterite (Mg_2SiO_4) to capture the gas in a mineral carbonation process.
Using the data below, determine the required annual supply of the mineral.

Molar mass data (kg/kmol)

Mg: 24; Si: 28; O: 16; C: 12.

Solution

The basic reaction is:

$$\frac{1}{2}Mg_2SiO_4 + CO_2 \rightarrow MgCO_3 + \frac{1}{2}SiO_2$$

Molar mass of forsterite (Mg_2SiO_4) = $(2 \times 24) + 28 + (4 \times 16) = 140$ kg/kmol
Molar mass of carbon dioxide (CO_2) = $12 + (2 \times 16) = 44$ kg/kmol
Ratio of forsterite to carbon dioxide is $(0.5)140/44 = 1.59$.

Therefore, to capture 20 Mt of CO_2, the forsterite requirement is $1.59 \times 20 = 31.82$ Mtonnes.

6.8 Tutorial Problems

Use the following properties where appropriate:

Universal gas constant: R_o: 8.314 kJ/kmol K

Molar masses: CO_2: 44.01 kg/kmol N_2: 28 kg/kmol

Molar specific heat capacity at constant volume: CO_2: 28.9 kJ/kmol K N_2: 21 kJ/kmol K

Specific gas constant for CO_2: 188.9 J/kg K.

6.1 Predict the pressure (MPa) of carbon dioxide (CO_2) at a temperature of 40 °C and a specific volume of 0.011939 m³/kg using:
   (a) The perfect gas law.
   (b) Van der Waal's equation.
   (c) The Beattie–Bridgeman equation.
   (d) The Benedict–Webb–Rubin equation.
   If the tabulated value is 4 MPa, determine the percentage error for each equation of state.

[Answers: 4.955 MPa (+23.9%), 4.072 MPa (+1.8%), 3.981 MPa (−0.4%), 4.043 MPa (1.0%)]

6.2 A gas mixture comprises 10 kmol of CO_2 and 2 kmol of N_2. Determine the mass (kg) of each gas and the apparent gas constant (kJ/kg K) for the mixture.

[Answers: CO_2 – 440 kg, N_2 – 56 kg, 0.202 kJ/kg K]

6.3 A gas mixture comprises 7 kg of CO_2 and 2 kg of N_2. Determine the pressure fraction for each gas and the apparent molar mass of the mixture (kg/kmol).

[Answers: CO_2 – 0.69, N_2 – 0.31, 39.04 kg/kmol]

6.4 Determine the reversible work (kJ/kg_{CO_2}) associated with the removal of carbon dioxide having a molar fraction of 0.1 in a gas mixture at a temperature of 120 °C.

[Answer: 170.94 kJ/kg_{CO_2}]

6.5 Carbon dioxide at 1 bar pressure and 20 °C is to be compressed to 9 bar. Determine the work of compression (kJ/kg) if the compression is carried out by:
   (a) A single-stage isentropic process ($k = 1.287$).
   (b) A single-stage polytropic process ($n = 1.2$).
   (c) A single-stage isothermal process ($n = 1$).
   (d) A two-stage polytropic compression process ($n = 1.2$) with an intermediate pressure of 3 bar and perfect intercooling, i.e. the gas enters the second stage of the compression at 293 K.

[Answers: 156.2 kJ/kg, 146.9 kJ/kg, 121.9 kJ/kg, 133.5 kJ/kg]

6.6 Carbon dioxide is to be compressed from 1 bar to 16 bar in a two-stage process. What is the optimal intermediate pressure for this process?

[Answer: 4 bar]

6.7 It is proposed that elemental iron ($M_{Fe} = 56$ kg/kmol) is to be used as the oxygen carrier in a chemical looping combustion system using coal (modelled by elemental carbon $M_C = 12$ kg/kmol). Using the simple fuel reactor equation, determine the approximate mass of iron (tonnes) required to burn 1 tonne of coal.
Give a reason why the amount of iron required, in practice, would be *in excess* of that indicated by the calculation.

[Answer: 9.33 tonnes Fe, excess air]

6.8 Assume the production of carbohydrate (starch) by green plants is governed by the simplified equation below:

$$CO_2 + H_2O \rightarrow (CH_2O) + O_2$$

Determine approximately how much CO_2 (ktonne) would be absorbed in the manufacture of 1 ktonne of starch and how much O_2 (ktonne) would be evolved.

[Answers: 1.47 ktonne CO_2, 1.07 ktonne O_2]

6.9 It has been suggested that the mineral serpentine ($Mg_3Si_2O_5(OH)_4$) be used to capture the carbon dioxide from a power plant emitting 5 Mt per annum. Determine the amount of serpentine (Mt/annum) and the percentage breakdown by mass of the reaction's products.

[Answers: 10.45 Mt, 61.8% $MgCO_3$, 29.4% SiO_2, 8.8% H_2O approximately]

6.10 A rock stratum having an area of 250×10^6 m^2, a thickness of 150 m and a porosity of 20% is selected as a potential CO_2 store. The CO_2 storage gas density is to be 700 kg/m^3 and the storage efficiency is predicted to be 40%. Determine the storage capacity (kg) and what percentage of an annual global CO_2 emission of 36 Gtonne could be accommodated by the store.

[Answers: 2.1×10^{12} kg, 5.8%]

7

Pollution Dispersal

7.1   Overview

Nothing is perfect and no 'end of pipe' pollution arrestor is 100% effective in cleaning up a polluted air stream. Something is always released. What is important is the concentration exposure of the environment to the emission upon release.

The most common (and usually the cheapest) way of quickly reducing the concentration of a discharge to the atmosphere is to release the emission via a tall chimney, stack or flue and use the atmosphere's turbulent mixing effect to disperse and dilute the pollutant.

National and internationally agreed environmental regulations and targets will dictate the parameters of the flue design (for example, flue height) and operation, perhaps enforcing these by ground-level environmental monitoring.

This chapter will provide an introduction to the characteristics of the atmosphere and go on to describe the part it plays in environmental dilution and in modelling the release of emissions from stacks.

Learning Outcomes

- To be familiar with the atmosphere's structure and composition.
- To understand the theoretical atmospheric pressure/temperature relationship.
- To gain an understanding of temperature–elevation or *lapse rates* in the atmosphere and their effect on vertical motion.
- To be familiar with atmospheric stability classifications and their effect on point source emission physical behaviour.
- To be able to use simple dispersal modelling to predict emission concentrations downwind of a single point source.
- To be familiar with commonly used correlations for estimating emission plume rise.
- To appreciate the effect of buildings and the surrounding landscape on emission dispersal from stacks.
- To be aware of alternative expressions of emission concentration.

Conventional and Alternative Power Generation: Thermodynamics, Mitigation and Sustainability,
First Edition. Neil Packer and Tarik Al-Shemmeri.
© 2018 John Wiley & Sons Ltd. Published 2018 by John Wiley & Sons Ltd.

7.2 Atmospheric Behaviour

7.2.1 The Atmosphere

The atmosphere is a mixture of gases and water separating the planet from space. Its maximum thickness is of the order of 8000 km. The mass of dry gas is of the order of 5×10^{18} kg whilst the mass of water is about 1.3×10^{13} kg. Its interface with space is gradual, becoming indistinct at the outer edge. Its composition and density are, however, by no means constant throughout its thickness.

The atmosphere has a layered structure and many subdivisions but, crudely speaking, the portion above 50 km is termed the *upper atmosphere* and that below 50 km the *lower atmosphere*, as illustrated in Figure 7.1.

The lower atmosphere contains approximately 98% of the atmosphere's total mass and can be further divided into:

- The *troposphere* – a dense, turbulent layer closest to the earth comprising ~ 80% by mass of the earth's atmosphere. Its thickness varies from about 8 km at the poles to 18 km at the equator. Seasonal variations in thickness can also occur.
  An often-used average thickness value is 11 km.
  It contains ~ 90% of the planet's atmospheric water content and is home to the world's weather systems. The interface of the troposphere with its overlying layer is termed the *tropopause.*
- The *stratosphere* – a less dense, stable layer extending from the tropopause up to about 50 km above the earth's surface, containing little water vapour and relatively high ozone (O_3) concentrations. Atmospheric pressure in the stratosphere is typically one tenth of the value at sea level. The interface of the stratosphere with the upper atmosphere is termed the *stratopause.*

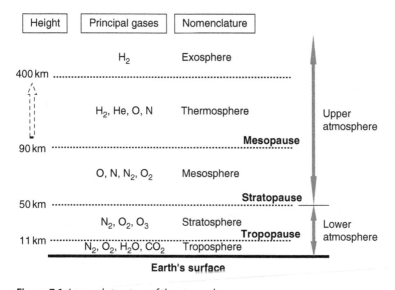

Figure 7.1 Layered structure of the atmosphere.

Table 7.1 Approximate composition of 'clean' air.

Component gas	Volume (%)	Concentration (ppm)
Nitrogen	78.09	780,900
Oxygen	20.94	209,400
Argon	0.93	9300
Carbon dioxide	0.04	~ 410 (May 2018)
Neon	trace	18.0
Helium	trace	5.2
Methane	trace	1.2
Krypton	trace	0.5
Hydrogen	trace	0.5

Source: Adapted from Williams, I. (2001) *Environmental Chemistry,* John Wiley & Sons.

The upper atmosphere contains the remaining 2% of the earth's atmosphere by mass and can be divided into:

- The *mesosphere* – a layer extending up to a height of about 90 km comprising mostly atomic oxygen and nitrogen. Atmospheric pressure in the mesosphere is typically 1/1000th of the value at sea level. The interface of the mesosphere with the overlying upper atmosphere is termed the *mesopause.*
- The *thermosphere* – a layer extending up to a height of about 400 km comprising mostly hydrogen and helium. Atmospheric pressure in the thermosphere is of the order of $1/10^{12}$th of the value at sea level. The thermosphere encompasses a highly ionized region (known as the ionosphere) responsible for the interception of much of the incoming short-wavelength solar and cosmic electromagnetic radiation.
- The *exosphere* – a layer extending from about 400 km above the planet to interplanetary space. The gas mixture in the exosphere is of a very low density and comprises mainly hydrogen. Gravitational attraction is low here and some atoms are able to escape to space.

As described earlier, the atmosphere is a mixture of gases ('dry air') and water vapour. The 'dry gas' comprises both chemically active gases that are essential for life and many inert gases. Most of these gases are colourless and odourless.

The discussion that follows in this section applies to the troposphere. An approximate breakdown by volume for '*pure, pollution-free*' dry air in the troposphere is given in Table 7.1.

7.2.2 Atmospheric Vertical Temperature Variation and Air Motion

In the atmosphere, any parcel of air that is less dense than its surroundings will rise due to its buoyancy.

Any parcel denser than its surroundings will sink due to negative buoyancy.

Hence, most localized *vertical* motions in the atmosphere are caused by changes in air density.

Note that on a global scale, horizontal motions are caused by warm air rising at the equator and sinking at the poles. This is further modified by the rotation of the earth about its axis.

The rate of temperature change with increasing atmospheric elevation or *atmospheric lapse rate* is of particular importance in pollutant-dilution studies, as it will affect the pollutant's ability to rise and disperse. This parameter can, in theory, be estimated with the application of some thermodynamics.

First, consider atmospheric air temperature changes with atmospheric pressure (dT/dP).

From a consideration of the first law of thermodynamics (non-flow) for an internally reversible process:

$$\delta q - \delta w = du \tag{7.1}$$

Assume a reversible and adiabatic (i.e. no heat transfer, $\delta q = 0$) condition and use the standard work ($\delta w = Pdv$) expression to give:

$$0 = du + Pdv \tag{7.2}$$

Remembering that the enthalpy of a substance can be described as the sum of its internal energy and any flow work, i.e. $h = u + Pv$:

$$dh = du + Pdv + vdP \tag{7.3}$$

Rearranging for du and substituting into the first law expression above:

$$0 = dh - vdP \tag{7.4}$$

Remembering that, for a gas at constant pressure, $dh = C_p dT$, substituting:

$$0 = C_p dT - vdP \tag{7.5}$$

Rearranging to give a temperature–pressure relationship:

$$\frac{dT}{dP} = \frac{v}{C_p} \tag{7.6}$$

Next, from a consideration of the fluid-static law, the relationship for a change in atmospheric pressure with elevation z is:

$$\frac{dP}{dz} = -\rho g \tag{7.7}$$

Note the negative sign as atmospheric pressure decreases with increasing elevation.

Finally, combining the above produces an expression for atmospheric air temperature change with elevation (dT/dz):

$$\frac{dT}{dz} = \frac{dT}{dP} \times \frac{dP}{dz} \tag{7.8}$$

Then

$$\frac{dT}{dz} = \frac{v}{C_p} \times (-\rho g) = -\frac{g}{C_p} \tag{7.9}$$

This temperature gradient is known as the *adiabatic lapse rate* (°C/m).

Using a typical average value for lower atmospheric air ($C_p = 1005$ J/kg K) renders a value of around −0.0097 °C/m or −9.7 °C/km.

A little care should be taken with the parameter, as both *dry* (moisture-free) air and *wet* lapse rates are quoted in the literature.

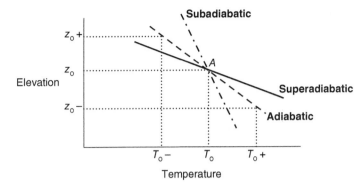

Figure 7.2 Temperature–elevation relationships.

If, under a given set of conditions, the numerical value of the actual ambient lapse rate exceeds the adiabatic rate, i.e. $(dT/dz)_{act} > (dT/dz)_{ad}$, the local atmospheric conditions are termed *superadiabatic*.

If, under a given set of conditions, the numerical value of the actual ambient lapse rate is less than the adiabatic rate, i.e. $(dT/dz)_{act} < (dT/dz)_{ad}$, the local atmospheric conditions are termed *subadiabatic*.

The inter-relationship between these three conditions is illustrated in Figure 7.2. Note that although the lapse rate is a temperature–elevation gradient, it is usually shown on a reversed axis arrangement with elevation on the vertical axis.

The presence of atmospheric water undergoing phase change and attendant heat transfer greatly complicates matters, and typical average tropospheric lapse rates are of the order of one third less than that indicated by theory.

Furthermore, adiabatic theory would indicate a temperature of around −93 °C at the tropopause. The actual temperature (∼ −57 °C) differs from that predicted by the theory due to infra-red heating from the earth's surface and the upward transfer of heat in cloud formation.

The lapse rate prediction breaks down completely in the stratosphere due to chemical reactions and the absorption of incoming solar energy.

7.3 Atmospheric Stability

Consider a parcel of air at point A (T_0, z_0) on Figure 7.2 as a starting point. Let a gust of wind initiate vertical motion.

If this happens rapidly enough, the parcel of air will follow the adiabatic line.

Relative to A, rising in the atmosphere will result in a reduction in temperature to (T_0-) and falling will result in a rise in temperature to $(T_0 +)$.

Its fate, however, will depend on the *lapse rate* conditions of its destination surroundings.

i) *Adiabatic surroundings* – Under this condition of equal density and temperature, the displaced air will not move any further under the influence of gravity. A condition of *neutral stability* is said to exist.

ii) *Subadiabatic surroundings* – If the initial displacement is upward to (z_0+) following the adiabatic line, it will be colder and denser than the surrounding subadiabatic air, thus negative buoyancy will cause it to sink. If the initial displacement is downward to $(z_0\ -)$, it will be warmer than the surrounding subadiabatic air and so buoyancy will cause it to rise. This condition is termed *stable*, and vertical air movement is inhibited.

iii) *Superadiabatic surroundings* – If the initial movement is upward to (z_0+) following the adiabatic curve, the air will be warmer and less dense than the surrounding superadiabatic air and buoyancy will cause it to continue to rise. If the initial displacement is downward to $(z_0\ -)$, the air will be colder and denser than the surrounding superadiabatic air and will sink. Under these circumstances, the atmosphere is termed *unstable* and vertical air motion is spontaneous.

A set of conditions of particular interest is the *inversion*. Under inversion conditions, the air temperature *increases* with elevation, yielding a very stable air mass which effectively becomes a pollutant trap eliminating vertical air motion.

Inversions can be produced by:

- Cooling a low-level air layer via radiation to the night sky.
- Heating a high-level air layer by solar radiation.
- A layer of warm air flowing over a colder layer, as can occur over snow fields.
- A layer of cold air flowing under a warm layer, as can occur down valley sides at night.

7.3.1 Stability Classifications

The previously described atmospheric stability phenomena were correlated with atmospheric conditions, i.e. solar radiation, wind speed and cloud cover by British meteorologist Frank Pasquil to produce the commonly used *stability classes* (A–G), as shown in Figure 7.3 and Table 7.2.

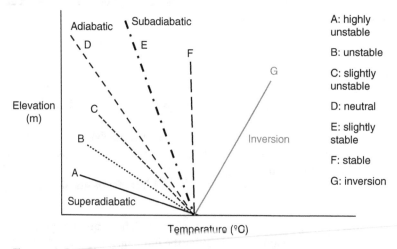

Figure 7.3 Pasquil stability temperature–elevation gradients.

Table 7.2 Pasquil climatological stability classifications.

Wind [a](m/s)	Solar radiation			Night time Cloud cover fraction	
	Strong [b]	Moderate [c]	Weak	>0.5 [d]	<0.375 [e]
<2	A	A–B [f]	B	E	F
2–3	A–B	B	C	E	F
3–5	B	B–C	C	D	E
5–6	C	C–D	D	D	D
>6	C	D	D	D	D

a) measured at 10 m height.
b) clear summer day with sun more than 60° above horizon.
c) summer day with a few broken clouds or a clear day with the sun 35–60° above the horizon.
d) fall afternoon or cloudy summer day with sun 15–35° above the horizon.
e) fractional cloud cover.
f) take average value of two dispersion classes.
Always use class D for overcast conditions.
Source. de Visscher, A. (2013) *Air Dispersion Modelling: Foundations and Applications*, John Wiley & Sons.

7.3.2 Stability and Stack Dispersal

If the ambient temperature profile is known and compared to the adiabatic lapse rate, it is possible to describe qualitatively the instantaneous stack emission plume behaviour.

7.3.2.1 Non-inversion Conditions

Three cases are illustrated in Figure 7.4.

In (a), an unstable atmosphere or superadiabatic lapse rate produces a plume that is characterized by *looping* in the longitudinal direction.

In (b), a neutral ambient lapse rate produces a *coning* plume in the longitudinal direction.

Figure 7.4 Stack plume behaviour for non-inversion conditions.

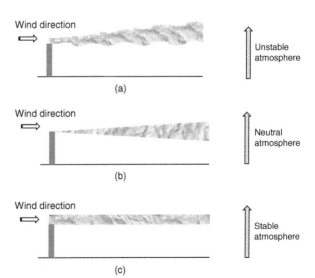

In (c), a stable ambient lapse rate produces a *fan-shaped* plume in the longitudinal direction with no vertical dispersal.

7.3.2.2 Inversion Conditions

These are illustrated in Figure 7.5.

Case (d) indicates the condition of *lofting* with a stable inversion below the stack tip and neutral conditions above. There is some mixing above the stack but none below.

In (e), near-neutral conditions below the stack tip and an inversion above inhibit mixing above the stack in the vertical direction, whilst mixing below results in the plume impacting the ground downwind. This combination of conditions is often termed *trapping*.

Case (f) indicates the condition for complete neutral stability below the stack tip and an inversion above. This set of conditions is often termed *fumigating*.

7.3.3 Variation in Wind Velocity with Elevation

Wind velocity has a significant influence on emission dispersal and so knowledge of its value at the stack tip is important.

Due to friction effects, wind speed varies with height from zero at ground level to a maximum at some height above the influence of buildings and local topography.

For very tall stacks, using ground-level wind velocity values might be misleading.

Wind speeds ($\bar{V}_{10m}$) are commonly monitored at a height of 10 m. The wind speed at any other height, z (m) can be determined from:

$$\bar{V}_z = \bar{V}_{10m} \left(\frac{z}{10} \right)^p \tag{7.10}$$

Typical values of exponent p for stability classifications and local terrain are shown in Table 7.3.

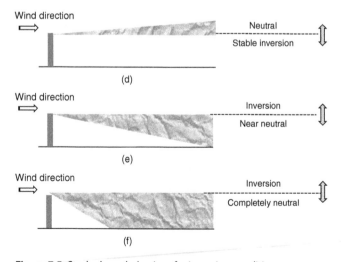

Figure 7.5 Stack plume behaviour for inversion conditions.

Table 7.3 Height-dependent wind velocity empirical parameters.

Stability class	Terrain	
	Smooth, rural	Urban
A	0.11	0.15
B	0.12	0.15
C	0.12	0.20
D	0.17	0.25
E	0.29	0.4
F	0.45	0.6

Source: de Visscher, A. (2013) *Air Dispersion Modelling: Foundations and Applications*, John Wiley & Sons.

7.4 Dispersion Modelling

7.4.1 Point Source Modelling

Consider an emission plume exiting a stack or chimney. At the stack exit, the plume cross-section can be considered to be circular with a radius equal to half that of the stack's inside diameter. Upon leaving the stack and contacting the atmosphere, the plume will be seen to bend over and travel in the direction of the prevailing wind.

Before assuming its horizontal path, the stack plume may experience some vertical displacement due to the presence of any inherent flow momentum and/or buoyancy force resulting from a plume temperature in excess of ambient conditions.

Once fully captured by the wind, the plume will normally start to experience spreading or dispersion in both the horizontal (y) and vertical (z) co-ordinates downwind of the stack, i.e. the x co-ordinate. This spreading is due to thermal and mechanical turbulent mixing and diffusion with the atmosphere that results in a reduction in emission concentration and a change in emission concentration distribution with increasing distance from the stack.

At any given distance downwind of the stack, it is assumed that the concentration variation in both vertical and horizontal planes can be adequately modelled by a bell-shaped or normal probability distribution curve, leading to the idea of a *Gaussian plume* dispersion model.

Single point source dispersion or diffusion models predict the average downwind concentration (kg/m³) in three co-ordinates (x, y, z) resulting from an emission. The usual model parameter arrangement is presented in Figure 7.6.

The model makes the following important assumptions:

- The source is located in a uniform terrain, e.g. open, flat country.
- The flue or chimney can be regarded as a small area or point source with a constant emission rate.
- There is no variation in wind speed between the source and the receptor.
- There is no variation in wind direction with height.
- The plume is non-reactive.

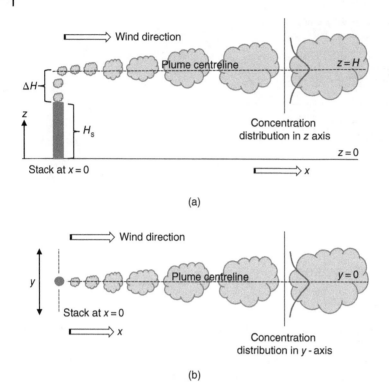

Figure 7.6 Gaussian plume dispersion spatial parameters. (a) Ground-level (elevation) view; (b) aerial (plan) view.

- No wet or dry deposition (fallout or sedimentation) occurs. This assumption is acceptable for emissions containing particulates with a diameter of less than 20 μm.
- Any plume impacting the ground is reflected.
- The emission point is assumed to be at an effective emission height, H, which is the sum of the stack height, H_{st} and any plume rise ΔH due to buoyancy and momentum effects, i.e.

$$H = H_{st} + \Delta H \qquad (7.11)$$

Intuitively, the concentration at any point in the plume will be directly proportional to the emission mass flow rate (kg/s) and indirectly proportional to the prevailing wind speed (m/s).

A full mathematical analysis is beyond the scope of this text, but simple point source concentration (c, kg/m^3) models are usually presented in a form that mimics the Gaussian probability function in both y and z co-ordinates, as shown below:

$$c_{(x,y,z)} = \frac{\dot{m}_e}{\bar{V}_w} \times \left\{ \frac{1}{(2\pi)^{0.5}\sigma_y} \exp\left[-\frac{1}{2}\left(\frac{y}{\sigma_y}\right)^2\right] \right\} \times \left\{ \frac{1}{(2\pi)^{0.5}\sigma_z} \exp\left[-\frac{1}{2}\left(\frac{z}{\sigma_z}\right)^2\right] \right\}$$

$$= \frac{\dot{m}_e}{2\pi\sigma_y\sigma_z\bar{V}_w} \times \exp\left[-\frac{1}{2}\left(\frac{y}{\sigma_y}\right)^2\right] \times \exp\left[-\frac{1}{2}\left(\frac{z}{\sigma_z}\right)^2\right] \qquad (7.12)$$

where σ_y and σ_z are modified diffusivity factors known as *dispersion coefficients* (m) in the y and z co-ordinates, respectively, which define the concentration distribution.

The first term in the product describes the centre line concentration of the plume at any distance downwind of the stack, and the second and third terms can be thought of as horizontal and vertical modifications to this value for the actual co-ordinates of interest, relative to the centre line. As the maximum concentration occurs at the centre line, the modifications have numerical values less than or equal to 1.

Concentration distributions in the y and z directions about the centre line are symmetrical according to the Gaussian form.

If the effective emission height is H (i.e. stack height + plume rise) then the z co-ordinate of the plume centre line is taken as $(z - H)$.

Substituting in the above equation and adding a $(z + H)$ term for plume reflection from the ground gives:

$$
c_{(x,y,z)} = \frac{\dot{m}_e}{2\pi\sigma_y\sigma_z\bar{V}_w} \times \exp\left[-\frac{1}{2}\left(\frac{y}{\sigma_y}\right)^2\right]
$$
$$
\times \left\{ \exp\left[-\frac{1}{2}\left(\frac{z-H}{\sigma_z}\right)^2\right] + \exp\left[-\frac{1}{2}\left(\frac{z+H}{\sigma_z}\right)^2\right] \right\} \tag{7.13}
$$

Further exponential terms can be added to accommodate the occurrence of inversion conditions.

Simplifications of the general form based on circumstances are common.

From the above, when interested in the *concentration in the plume at its centre line*, i.e. $z = H$, $y = 0$, then the above reduces to:

$$
c_{(x,0,0)} = \frac{\dot{m}_e}{2\pi\sigma_y\sigma_z\bar{V}_w} \tag{7.14}
$$

Ground-level concentrations (i.e. $z = 0$) are usually of the greatest interest. Under this circumstance, the general form reduces to:

$$
c_{(x,y,0)} = \frac{\dot{m}_e}{\pi\sigma_y\sigma_z\bar{V}_w} \exp\left[-\frac{1}{2}\left(\frac{y}{\sigma_y}\right)^2\right] \exp\left[-\frac{1}{2}\left(\frac{H}{\sigma_z}\right)^2\right] \tag{7.15}
$$

When interested in the *ground-level concentration* occurring *under the plume centre line* ($y = 0$, $z = 0$), the general form [remembering that exp (0) = 1] reduces to:

$$
c_{(x,0,0)} = \frac{\dot{m}_e}{\pi\sigma_y\sigma_z\bar{V}_w} \exp\left[-\frac{1}{2}\left(\frac{H}{\sigma_z}\right)^2\right] \tag{7.16}
$$

For ground-level concentration resulting from a *ground-level emission source* ($z = 0$, $H = 0$), as, for example, in the case of a dusty, uncovered solid fuel store, the general concentration equation becomes:

$$
c_{(x,y,0)} = \frac{\dot{m}_e}{\pi\sigma_y\sigma_z\bar{V}_w} \exp\left[-\frac{1}{2}\left(\frac{y}{\sigma_y}\right)^2\right] \tag{7.17}
$$

For ground-level concentration *under the plume centre line from* a *ground-level emission source* ($y = 0, z = 0, H = 0$), the general concentration equation is simplified to:

$$c_{(x,0,0)} = \frac{\dot{m}_e}{\pi\sigma_y\sigma_z\bar{V}_w} \tag{7.18}$$

Values of dispersion coefficients σ_y and σ_z as a function of atmospheric stability and distance downwind of the source have been produced through observation and measurement in open or rural country and urban conditions.

The σ values are usually presented in a graphical form plotted against distance downwind from source (x, m) and prevailing atmospheric stability conditions (A, B, C, etc.), as shown in Figures 7.7 and 7.8 for 'rural' conditions.

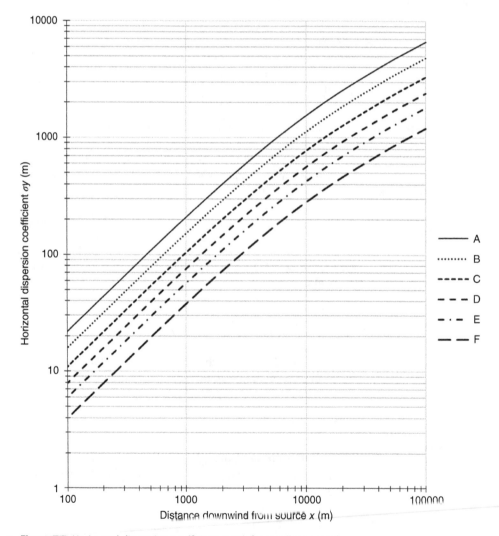

Figure 7.7 Horizontal dispersion coefficient graph for 'rural' topography.

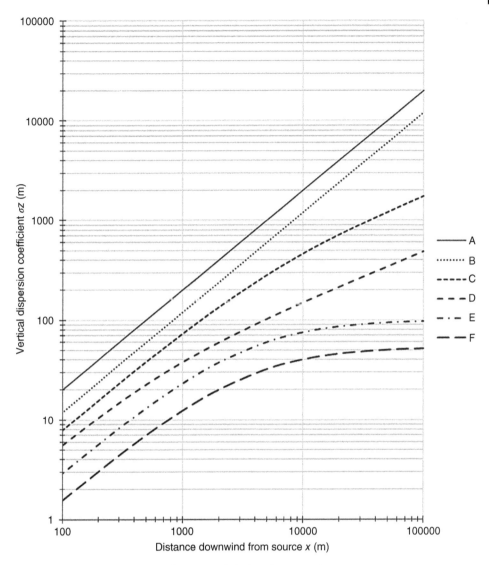

Figure 7.8 Vertical dispersion coefficient graph for 'rural' topography.

It is important to note the following:

- The data are presented on a log–log basis.
- Dispersion coefficient values increase with increasing distance downwind of the source.
- For any given distance downwind and stability class, the value of the horizontal dispersion coefficient generally exceeds the value of the vertical coefficient, meaning that a concentration contour at a given distance downwind would be an ellipse flattened in the vertical axis.
- As a consequence of buoyancy forces, at any given distance downwind, the spread of values is greater for the vertical dispersion coefficient.

Table 7.4 Dispersion coefficient correlations suitable for a downwind range (x) of 100–10 000 m.

a) Rural

Stability class	σ_y (m)	σ_z (m)
A	$0.22x(1 + 0.0001x)^{-0.5}$	$0.2x$
B	$0.16x(1 + 0.0001x)^{-0.5}$	$0.12x$
C	$0.11x(1 + 0.0001x)^{-0.5}$	$0.08x(1 + 0.0002x)^{-0.5}$
D	$0.08x(1 + 0.0001x)^{-0.5}$	$0.06x(1 + 0.0015x)^{-0.5}$
E	$0.06x(1 + 0.0001x)^{-0.5}$	$0.03x(1 + 0.0003x)^{-1}$
F	$0.04x(1 + 0.0001x)^{-0.5}$	$0.016x(1 + 0.0003x)^{-1}$

b) Urban

Stability class	σ_y (m)	σ_z (m)
A–B	$0.32x(1 + 0.0004x)^{-0.5}$	$0.24x(1 + 0.001x)^{-0.5}$
C	$0.22x(1 + 0.0004x)^{-0.5}$	$0.2x$
D	$0.16x(1 + 0.0004x)^{-0.5}$	$0.14x(1 + 0.0003x)^{-0.5}$
E–F	$0.11x(1 + 0.0004x)^{-0.5}$	$0.08x(1 + 0.00015x)^{-0.5}$

Source: de Visscher, A. (2013) *Air Dispersion Modelling: Foundations and Applications,* John Wiley & Sons.

Alternatively, correlation curve-fitting equations can be used. A commonly quoted set of curve-fitting equations (known as the *Briggs sigmas* after their progenitor, G.A. Briggs) derived in the late 1960s and early 1970s is presented in Table 7.4 for both 'rural' and 'urban' conditions.

Other correlations, which provide more detailed outputs depending on ranges of downwind distances and conditions, are available.

7.4.2 Plume Rise

An emission plume may rise some vertical distance before being entrained by the prevailing wind direction and, as mentioned earlier, the effective emission height (H) will therefore be dependent on stack height (H_{st}) plus any plume rise (ΔH) due to momentum and buoyancy.

The effective plume rise is estimated using a range of predictive formulae.

A reasonably simple experimental formula in common usage for power plant is the *Holland formula*:

$$\Delta H = \frac{2r_{st}\bar{V}_e}{\bar{V}_w}\left[1.5 + \left(2.68 \times 10^{-5}P\left(\frac{T_e - T_{amb}}{T_e}\right)2r_{st}\right)\right] \times CF \tag{7.19}$$

where CF is a correction factor (0.8–1.2) based on the Pasquil stability classification; r_{st} is the inside stack radius (m) and P is atmospheric pressure (Pa). Both emission and ambient temperatures are evaluated in Kelvin.

More complex empirical equations requiring an algorithmic approach for estimating plume rise (ΔH) are the *Briggs formulae*.

Using the Briggs formulae depends on the definition and estimation of some empirical parameters:

- Buoyancy flux parameter, BF (m^4/s^3):

$$BF = g\bar{V}_e r_{st}^2 ((T_e - T_{amb})/T_e) \qquad (7.20)$$

- x_{max}: the distance (m) downwind of the source to the maximum plume rise:

  If $BF < 55$ then $x_{max} = 50\,BF^{5/8}$ $\qquad (7.21)$

  If $BF \geq 55$ then $x_{max} = 120\,BF^{0.4}$ $\qquad (7.22)$

- Stability parameter, SP:

$$SP = (g/T_{amb}) \times ((\Delta T_{amb}/\Delta z) + 0.01) \qquad (7.23)$$

Then

1. For stability classes A–D:

$$\Delta H = (1.6BF^{0.333}x_{max}^{0.666})/\bar{V}_w \qquad (7.24)$$

[Note if x is less than x_{max} then it can be substituted directly into the plume rise equation above.]

2. For stability classes E–F:

$$\Delta H = 2.4(BF/\bar{V}_w SP)^{0.33} \qquad (7.25)$$

Correlations are also available to account for significant momentum-sourced plume rise; however, power plant calculations are usually limited to buoyancy assessments.

7.4.3 Effect of Non-uniform Terrain on Dispersal

It is rare that emission stacks are located in open, flat country. Often, dispersal has to contend with the presence of buildings and an undulating landscape.

Due to differences in radiant heat exchange or emissivity, even the colour or texture of flat, open land can affect air flow patterns, for example, dark-coloured surfaces like ploughed fields may absorb solar radiation better than more reflective surfaces, resulting in localized air heating and vertical air movement.

The presence of valleys and hills can modify ambient local wind directions and flow velocities, resulting in funnelling. Differential solar heating of slopes can result in localized convective (upward and downward) air flow. At night, radiation to the sky from exposed land surfaces can result in localized air cooling and collection in topographical depressions.

Localities near the sea are also cause for concern. Due to the different thermal capacity characteristics of the land and sea, in response to solar heating, a locality may experience air movement from sea to land during the day and the reverse at night.

Tall buildings adjacent to a stack affect flow patterns and interfere with dispersal, introducing vortices, recirculating flows, flow separation and stagnation points on the building itself.

Ultimately, detailed modelling should be carried out in all of the above cases to determine any unwanted/unhelpful effects.

7.5 Alternative Expressions of Concentration

Throughout the analysis in this chapter, the units of emission concentration have been expressed in SI terms, i.e. kg/m^3 or one of its submultiples.

However, other forms of expression are commonly encountered in the field of emission dispersal. One of the most common is the practice of expressing concentration in simple fraction or 'parts per' terms, abbreviated to 'pp', resulting in such terms as *parts per million* (ppm) or *parts per billion* (ppb).

A further complication is introduced when considering whether this relates to a volume fraction or a mass fraction. In general, solid pollutant concentrations are usually expressed in mass terms and gas concentrations in volume terms. To differentiate, a gaseous concentration is best expressed as 'ppmv' or 'ppbv'.

The SI equivalent (kg/m^3) of a fractional gas concentration (ppbv) can be estimated with knowledge of the specific gas constant (R, J/kg K), pressure (P, Pa) and temperature (T, K) of the gas, as follows:

$$c(kg/m^3) = c(ppbv) \times 10^{-9} \times \frac{P}{R \times T} \tag{7.26}$$

The gas constant R can be calculated from knowledge of the molecular mass (M, kg/kmol) of the gas of interest, as follows:

$$R = 1000 \times R_o/M \tag{7.27}$$

where R_o is the universal gas constant (kJ/kmol K).

7.6 Worked Examples

Worked Example 7.1 – Calculating plume concentration
A stack discharge of 70 g/s has an effective emission height of 175 m above ground level. The ambient wind velocity is 4 m/s at the stack tip. The horizontal and vertical dispersion coefficients at a point of interest downwind have been estimated to be 80 m and 38 m, respectively.

Determine, at the distance downwind of interest:

(a) The plume centre line concentration ($\mu g/m^3$).
(b) The concentration ($\mu g/m^3$) at plume height but horizontally offset 100 m from the plume centre line.
(c) The concentration ($\mu g/m^3$) 50 m directly below the plume centre line.

Solution
Given: $\dot{m}_e - 70 \times 10^{-3}$ kg/s, $\bar{V}_w - 4m/s$, $H = 175$ m, $\sigma_y = 80$ m, $\sigma_z = 38$ m.
Find c.

(a) Plume centre line, i.e. $y = 0, z = H$

$$c_{(x,y,z)} = \frac{\dot{m}_e}{2\pi\sigma_y\sigma_z\bar{V}_w} \times \exp\left[-\frac{1}{2}\left(\frac{y}{\sigma_y}\right)^2\right]$$

$$\times \left\{\exp\left[-\frac{1}{2}\left(\frac{z-H}{\sigma_z}\right)^2\right] + \exp\left[-\frac{1}{2}\left(\frac{z+H}{\sigma_z}\right)^2\right]\right\}$$

$$= \frac{0.07}{2\pi \times 80 \times 38 \times 4} \times \exp\left[-\frac{1}{2}\left(\frac{0}{80}\right)^2\right]$$

$$\times \left\{\exp\left[-\frac{1}{2}\left(\frac{0}{38}\right)^2\right] + \exp\left[-\frac{1}{2}\left(\frac{350}{38}\right)^2\right]\right\}$$

$$= 9.16 \times 10^{-7} \times 1 \times \{(\cong 1)\} = 9.16 \times 10^{-7} \text{ kg/m}^3 = 916 \text{ µg/m}^3.$$

(b) Plume centre line height but 100 m horizontally offset, i.e. $y = 100$ m, $z = H$

$$c_{(x,y,z)} = \frac{\dot{m}_e}{2\pi\sigma_y\sigma_z\bar{V}_w} \times \exp\left[-\frac{1}{2}\left(\frac{y}{\sigma_y}\right)^2\right]$$

$$\times \left\{\exp\left[-\frac{1}{2}\left(\frac{z-H}{\sigma_z}\right)^2\right] + \exp\left[-\frac{1}{2}\left(\frac{z+H}{\sigma_z}\right)^2\right]\right\}$$

$$= \frac{\dot{m}_e}{2\pi\sigma_y\sigma_z\bar{V}_w} \times \exp\left[-\frac{1}{2}\left(\frac{100}{80}\right)^2\right]$$

$$\times \left\{\exp\left[-\frac{1}{2}\left(\frac{0}{38}\right)^2\right] + \exp\left[-\frac{1}{2}\left(\frac{350}{38}\right)^2\right]\right\}$$

$$= 9.16 \times 10^{-7} \times 0.458 \times \{\cong 1\} = 4.19 \times 10^{-7} \text{ kg/m}^3 = 419 \text{ µg/m}^3.$$

(c) 50 m directly below plume centre line, i.e. $y = 0, z = 175 - 50 = 125$ m

$$c_{(x,y,z)} = \frac{\dot{m}_e}{2\pi\sigma_y\sigma_z\bar{V}_w} \times \exp\left[-\frac{1}{2}\left(\frac{y}{\sigma_y}\right)^2\right]$$

$$\times \left\{\exp\left[-\frac{1}{2}\left(\frac{z-H}{\sigma_z}\right)^2\right] + \exp\left[-\frac{1}{2}\left(\frac{z+H}{\sigma_z}\right)^2\right]\right\}$$

$$= \frac{0.07}{2\pi \times 80 \times 38 \times 4} \times \exp\left[-\frac{1}{2}\left(\frac{0}{80}\right)^2\right]$$

$$\times \left\{\exp\left[-\frac{1}{2}\left(\frac{-50}{38}\right)^2\right] + \exp\left[-\frac{1}{2}\left(\frac{300}{38}\right)^2\right]\right\}$$

$$= 9.16 \times 10^{-7} \times 1 \times \{\cong 0.42\} = 3.85 \times 10^{-7} \text{ kg/m}^3 = 385 \text{ µg/m}^3.$$

(Note that, in this example, the contribution from the 'reflected plume' modification is negligible; hence, the approximation in the third term.)

Worked Example 7.2 – Calculating ground-level concentration

A stack in open country has a height of 30 m with zero plume rise and emits 100 g/s of nitrogen oxides.

On an overcast day, the wind velocity at 10 m above ground is 4 m/s.

Determine the ground-level concentration (kg/m^3) at a point 2 km directly downwind of the stack.

Solution

Given: $\dot{m}_e = 100 \times 10^{-3}$ kg/s, $x = 2000$ m, $z = 0$, $y = 0$, $\bar{V}_{10m} = 4$ m/s, $H_{st} = 30$ m, $\Delta H = 0$.

Weak solar radiation (*overcast day*) – suggests Pasquil stability class C.

Open country – suggests wind velocity empirical exponent $p = 0.12$.

Find $c_{(2000,0,0)}$.

$$\bar{V}_z = \bar{V}_{10m} \left(\frac{z}{10}\right)^p = 4\left(\frac{30}{10}\right)^{0.12} = 4.56 \text{ m/s}$$

$$\sigma_y = 0.11x \times (1 + 0.0001x)^{-0.5} = 201 \text{ m}$$

$$\sigma_z = 0.08x \times (1 + 0.0002x)^{-0.5} = 135 \text{ m}$$

$$c_{(x,y,z)} = \frac{\dot{m}_e}{\pi \sigma_y \sigma_z \bar{V}_w} \times \exp\left[-\frac{1}{2}\left(\frac{y}{\sigma_y}\right)^2\right] \times \exp\left[-\frac{1}{2}\left(\frac{H}{\sigma_z}\right)^2\right]$$

$$c_{(2000,0,0)} = \frac{100 \times 10^{-3}}{\pi \times 201 \times 135 \times 4.56} \times [1] \times \exp\left[-\frac{1}{2}\left(\frac{30}{135}\right)^2\right]$$

$$= 251 \times 10^{-9} \text{ kg/m}^3$$

$$= 251 \text{ µg/m}^3.$$

Worked Example 7.3 – Determination of stack height

Using the data below, determine the stack height (H_{st}) for a fossil fuel power plant emitting 530 g/s of total suspended particulates if the ambient ground-level limit is to be limited to 75 µg/m³ at a site 5 km downwind of the source.

Take the design atmospheric conditions to be neutral (D) and a rural dispersion model.

Data

Gas exit velocity: 15 m/s; gas exit temperature: 150 °C; stack tip diameter: 3 m
Ambient temperature: 20 °C; prevailing wind velocity: 6 m/s.

Solution

Given: $\dot{m}_e = 530 \times 10^{-3}$ kg/s, $c = 260 \times 10^{-9}$ kg/m³, $x = 5000$ m, $\bar{V}_e = 15$ m/s, $T_e = 423$ K, $T_a = 293$ K, $r_{st} = 1.5$ m, $\bar{V}_w = 6$ m/s, stability class D.
Find H_{st}.

1. Determine dispersion coefficients from Pasquil–Gifford charts:

$$\sigma_y = 0.08x \times (1 + 0.0001x)^{-0.5} = 327 \text{ m}$$

$$\sigma_z = 0.06x \times (1 + 0.0015x)^{-0.5} = 103 \text{ m}$$

2. Determine the effective stack height (H) from the appropriate concentration equation:

$$c_{(5000, 0, 0)} = \frac{\dot{m}_e}{\pi \sigma_y \sigma_z \bar{V}_w} \exp\left[-\frac{1}{2}\left(\frac{H}{\sigma_z}\right)^2\right]$$

$$75 \times 10^{-9} = \frac{530 \times 10^{-3}}{\pi \times 327 \times 103 \times 6} \exp\left[-\frac{1}{2}\left(\frac{H}{103}\right)^2\right]$$

$$H = 226 \text{ m}$$

3. Select appropriate plume rise correlation:

$$BF = g\bar{V}_e r_{st}^2 (T_e - T_a)/T_e$$

$$= 9.81 \times 15 \times 1.5^2 \times (423 - 293)/423 = 102 \text{ m}^4/\text{s}^3, x_{max} = 120BF^{0.4} = 763 \text{ m}$$

$$\Delta H = \frac{1.6BF^{0.333} x_{max}^{0.666}}{\bar{V}_w} = \frac{1.6 \times 102^{0.333} \times 763^{0.666}}{6} = 103 \text{ m}$$

4. Determine stack height (H_{st})

$$H = H_{st} + \Delta H$$

$$226 = H_{st} + 103$$

$$H_{st} = 123 \text{ m}.$$

Worked Example 7.4 – Concentration conversion

A sulphur dioxide (SO_2) emission concentration is reported to be 200 ppbv.

If the reporting temperature and pressure are 20 °C and 1.013 bar, respectively, use the data below to express this concentration in $\mu g/m^3$.

Data

Molar mass of sulphur: 32 kg/kmol
Molar mass of oxygen: 16 kg/kmol
Universal gas constant: 8.314 kJ/kmol K.

Solution

Given: $c = 200$ ppbv, $T = 20$ °C $= 293$ K, $P = 1.013$ bar $= 1.013 \times 10^5$ Pa, $M_{SO_2} = 32 + 2(16) = 64$ kg/kmol.

Find c.

$$R = R_o/M = 1000 \times 8.314/64 = 130 \text{ J/kg K}$$

$$c(\text{kg/m}^3) = c(\text{ppbv}) \times 10^{-9} \times \frac{P}{R \times T} = 200 \times 10^{-9} \times \frac{1.013 \times 10^5}{130 \times 293}$$

$$= 5.32 \times 10^{-7} \text{ kg/m}^3 = 532 \text{ }\mu g/m^3.$$

7.7 Tutorial Problems

Note: Answers have been calculated using the curve-fit correlations supplied in the text. Use of dispersion coefficient charts is permitted, but numerically identical results should not be expected, as estimating from a graph is not an exact science.

7.1 The wind velocity at 10 m above ground on a totally overcast day in an urban location is determined to be 3.5 m/s.

Determine the wind velocity (m/s) 30 m above ground at the location.

[Answer: ~ 4.4 m/s]

7.2 Determine the prevailing stability classification and the horizontal and vertical dispersion coefficients (m) at a point 2500 m downwind of an emission source in rural surroundings on a moderately sunny day when the wind velocity at the plume centre line is 2.5 m/s.

What likely form will the plume profile take?

[Answers: B, 358 m, 300 m, looping]

7.3 A stack discharge of 800 g/s has an effective emission height of 200 m above ground level in an ambient wind velocity of 5 m/s at the stack tip. The horizontal and vertical dispersion coefficients at a point of interest downwind have been estimated to be 320 m and 100 m, respectively.

Determine, at the distance downwind of interest:

(a) The plume centre line concentration ($\mu g/m^3$).

(b) The concentration ($\mu g/m^3$) at plume height but horizontally offset 150 m from the plume centre line.

(c) The concentration ($\mu g/m^3$) 150 m directly below the plume centre line.

[Answers: 796 $\mu g/m^3$, 712 $\mu g/m^3$, 258 $\mu g/m^3$]

7.4 A process plant is emitting 500 g/s of particulates from a stack 125 m high with a plume rise of 75 m. The wind speed at the stack tip is 6 m/s and the stability category is C. The local environment is assessed to be 'urban'.

Determine the *maximum* ground-level concentration ($\mu g/m^3$) under the plume centre line and its position (m) downwind of the plant.

(*Suggestion* – Guess some values of downwind distance x and plot with a spreadsheet.)

[Answers: ~ 500 $\mu g/m^3$, ~ 750 m]

7.5 On a windy day, an uncovered solid fuel store is releasing particulates at an unknown rate. At a distance of 1000 m *directly downwind* of the source, the measured ground-level source of the pollutant is 75 $\mu g/m^3$.

If the prevailing stability class is D, the countryside is assessed to be 'open' and the prevailing wind velocity is 6 m/s, determine the emission rate (kg/s) of the source.

[Answer: ~ 4.08 × 10^{-3} kg/s]

7.6 A power plant in a rural setting emits 400 g/s of SO_2. If the ground-level concentration at a point directly under the plume centre line 1 km downwind of the stack is to be limited to 550 $\mu g/m^3$, determine the necessary effective emission height (m) on a day when the atmospheric stability class is C and the wind velocity at the stack tip is 5.5 m/s.

[Answer: 135 m]

7.7 An emission leaves a 2 m diameter stack exit at a velocity of 12 m/s and temperature of 90 °C. Local atmospheric conditions at the stack exit are 1 bar, 10 °C and a wind velocity of 3.5 m/s. The atmospheric stability class is believed to be C ($CF = 1.05$).
Use the Holland formula to predict the plume rise (m) from the stack.

[Answer: ~ 19 m]

7.8 Use the data below to determine the plume rise for a fossil fuel power plant under *neutral* atmospheric conditions by applying the appropriate Briggs formula.

Data

Stack tip radius: 1.75 m; ambient temperature: 20 °C; stack exit temperature: 140 °C; exit gas velocity: 10 m/s; wind speed at stack tip: 6 m/s.

[Answer: 97 m]

7.9 A power plant stack is 90 m in height and the stack tip has a diameter of 2 m. The emission velocity at the stack tip is 9 m/s and has a temperature of 135 °C. The ambient temperature lapse rate is 0.02 K/m and the wind velocity at stack tip is 6 m/s. The local atmosphere is estimated to be *slightly stable*. The ambient air temperature is 10 °C. The stack emits 600 g/s of NO_2.
If the stack is sited in open country, determine the NO_x concentration ($\mu g/m^3$) directly under the plume at a point at ground level 1500 m downwind of the stack.

[Answer: 1486 $\mu g/m^3$]

7.10 A nitrogen dioxide (NO_2) emission concentration is reported to be 350 ppbv. If the reporting temperature and pressure are 25 °C and 1.01325 bar, respectively, use the data below to express this concentration in $\mu g/m^3$.

Data

Molar mass of nitrogen(N): 14 kg/kmol
Molar mass of oxygen(O): 16 kg/kmol
Universal gas constant: 8.314 kJ/kmol K.

[Answer: 658 $\mu g/m^3$]

8

Alternative Energy and Power Plants

8.1 Overview

Global warming is one of the most serious challenges facing humanity today. It has come about partly due to the consumption of electricity, partly due to the burning of fuel for heating and transport and partly due to the increase in global human and livestock populations. Its impacts adversely affect all aspects of the natural world. To protect the health and economic well-being of current and future generations, we must reduce our emissions of greenhouse gases by using the technology, know-how and practical solutions already at our disposal. In particular, it has been proposed that the world's electrical power demands from coal must fall and be replaced by a renewable fuel such as solar, wind, bioenergy or hydropower.

There is a unified international agreement to limit the use of fossil fuels and to adopt renewables to cover some of our demands for power and heating. For example, in Europe, the EU20-20 accord is an attempt to source 20% of energy consumption from renewables by the year 2020.

The recent and projected growth for global electrical energy generation (1990 to 2040), as predicted by the US Energy Information Administration, is shown in Figure 8.1.

This chapter describes some of the alternative, non-fossil-fuel energy technologies, including the non-renewable nuclear option, available to combat global warming and meet these objectives.

Learning Outcomes

- To gain an overview of non-fossil-fuel energy sources available for the production of electricity.
- To understand the practical applications of their related technologies.
- To be able to understand the theory behind these systems and estimate the power output from some typical units.
- To be able to solve problems related to alternative energy systems.

Conventional and Alternative Power Generation: Thermodynamics, Mitigation and Sustainability,
First Edition. Neil Packer and Tarik Al-Shemmeri.
© 2018 John Wiley & Sons Ltd. Published 2018 by John Wiley & Sons Ltd.

world electricity generation by fuel
billion kilowatthours

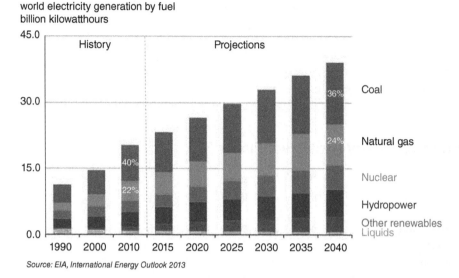

Source: EIA, International Energy Outlook 2013

Figure 8.1 World electricity generation 1990–2040. Data courtesy of EIA 2013 report.

8.2 Nuclear Power Plants

Currently, nuclear power produces approximately 17% of global electricity. The contribution varies from country to country. In France, over 75% of electricity is produced by nuclear power. The United States and the United Kingdom, on the other hand, only produce about a quarter of their electricity from nuclear fission.

Nuclear fission is the splitting of an atom's nucleus, by a neutron, into lighter nuclei, electrons and further neutrons and perhaps electromagnetic radiation.

Fission-sourced neutrons can be captured by other adjacent atoms to continue the (chain) reaction. In the presence of sufficient nuclear material, the reaction becomes self-sustaining.

Nuclear fission also produces heat as a by-product, and this heat can be used to make steam for use in a conventional power cycle. Hence, aside from the heat-generation method, nuclear power generates electricity much like a coal-fired power plant but without the associated carbon dioxide emissions. Like their fossil-fuel counterparts, nuclear power plants can also be a huge source of thermal energy for heating, if required.

Due to the difference in the binding energies of the parent atom and its fission products, the nuclear fission of a single uranium-235 nucleus can produce up to 200 MeV or 3.2×10^{-11} J of energy. This approximates to 82 TJ/kg and is quite impressive when compared with that of 34 MJ/kg for an anthracite coal.

8.2.1 Components of a Typical Nuclear Reactor

A nuclear reactor consists of a vessel with fuel rods, control rods, the moderator and a coolant.

Reactor fuel rods are commonly made of uranium oxide.

Control rods are made from a material (for example, boron, cadmium, hafnium) with good neutron-absorbing capabilities, useful in regulating or shutting down the reaction.

The moderator is a material (for example, graphite or water) surrounding the fuel rods that is used to *slow down* generated neutrons to a speed or energy conducive to capture by uranium nuclei.

The coolant is a fluid (for example, water, CO_2, liquid sodium) used to collect the thermal energy generated by the reactor.

A containment structure prevents radioactivity from escaping to the environment.

8.2.2 Types of Nuclear Reactor

Water reactors can be of the pressurized-water or boiling-water variety.

In a pressurized-water reactor, the thermal energy of the nuclear reaction is used to heat a primary coolant circuit that produces steam in a secondary circuit via an indirect heat exchanger (see Figure 8.2). The steam is then routed to the turbine set for power production.

In a boiling-water reactor, the indirect heat exchanger and secondary circuit are dispensed with and power is generated by sending steam directly to a turbine.

8.2.3 Environmental Impact of Nuclear Reactors

Although proponents of nuclear power often emphasize the carbon-free aspect of the technology, all forms of radiation, whether man-made or naturally occurring, are potentially dangerous, and therefore nuclear power is not without its own radioactive waste issues.

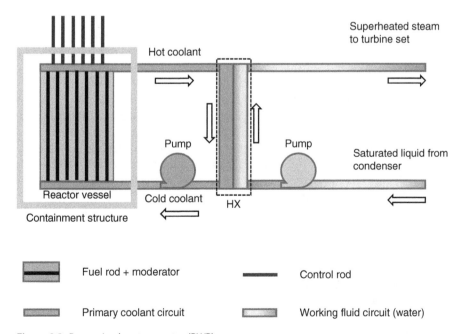

Figure 8.2 Pressurized-water reactor (PWR).

Nuclear power waste is commonly classified into:

- *High level* – spent reactor fuel, reprocessing material.
- *Intermediate level* – reactor ancillary equipment, e.g. control rods, instrumentation etc.
- *Low level* – contaminated laboratory material, operator clothing, miscellaneous rubbish.

During its passage through matter, radiation knocks electrons out of atoms or molecules (*ionization*). There is no known lower limit of radiation below which no damage is done to living tissue, and the human species is more sensitive than most to radioactivity. Some forms of radioactivity are persistent, having a high mobility in the environment and a tendency to concentrate in biological systems.

Radiation damage takes three forms: immediate physical damage to tissues ('radiation burns'); delayed damage (cancer); and genetic abnormalities (mutations) including cancer in successive generations.

There are three common SI units describing the effects of radioactive emission/absorption:

Becquerel – Quantifies the emission rate or activity of a radioactive source.

Gray – An absorbed dose of radioactive energy (J/kg of tissue).

Sievert – An absorbed dose modified by a factor to account for the relative susceptibility of the irradiated tissue to a particular type of ionizing radiation, e.g. alpha particles, beta particles, gamma rays etc.

8.3 Solar Power Plants

The sun is an enormous nuclear fusion reactor around which the earth orbits at a distance of approximately 150×10^9 m. The core of the sun is estimated to be at a temperature of some 10 million Kelvin whilst its outer surface is at a relatively cool 6000 K.

A little heat transfer theory indicates that the radiation power per square metre of surface area of a body is the product of a universal constant multiplied by the body's temperature (K) raised to the fourth power. Using this, some arithmetic shows that the sun emits at a rate of around 60–70 MW/m^2.

Assuming the sun to have a radius of around 0.7×10^9 m, the total power output for the sun is of the order of $3–4 \times 10^{20}$ MW. Thanks to a (distance-related) inverse square relationship, the resulting solar intensity at the earth's orbit is found to be only about 1372 W/m^2. This figure is termed the *solar constant*.

The amount of solar energy reaching the surface is reduced to below the solar constant by:

- Atmospheric reflection
- Atmospheric absorption by O_3, H_2O, O_2 and CO_2
- Rayleigh scattering (by small atmospheric particles)
- Mie scattering (by large atmospheric particles).

The relative extent of these effects is light wavelength/particle or molecule-dimension dependent.

The thickness of air (or air mass) between the incoming solar radiation and the surface will, of course, be important and will vary with sun height (or angle).

Nevertheless, even after atmospheric absorption and reflection, a useful amount of solar energy reaches the earth's surface.

The variation with latitude is significant. For example, the average annual incident solar irradiance in northern and central Europe is 700–1000 kWh/m². Southern Europe enjoys an average annual amount of sunshine in excess of 1700 kWh/m². Desert regions in northern Africa receive around 2500 kWh/m².annum, on average.

To put this in context, on average, the earth's surface receives about 1.2×10^{17} W of solar power. If it could be converted at 100% efficiency, the planet receives an amount of energy equal to its total annual energy demand in less than one hour.

Solar radiation reaches the surface in two forms: direct and diffuse. Direct (or shadow-casting) solar radiation is more intense, whereas diffuse (or directionally unde-fined) solar radiation, although carrying less energy, is still able to make a significant contribution to energy capture.

Both forms of solar radiation are utilized by two conversion technologies: solar pho-tovoltaic (PV) panels and solar thermal collectors.

Converting this huge potential into a useful form of output, such as heating or elec-tricity, has been somewhat less than proactive to date. However, the utilization of global solar energy for electrical and thermal applications has risen over the period 2005 to 2015, going from 5.1 to 227 GW for electricity and from 100 to 435 GW for solar heating, as shown in Figure 8.3.

8.3.1 Photovoltaic Power Plants

In 1839, French scientist Henri Becquerel discovered the photo effect; that is, that in some substances, electrons can be separated from their atoms by exposure to discrete packets of light or photons.

The phenomenon can be understood by considering electron orbits. The outermost band (or orbit) that an electron can occupy and still be closely associated with its atom is called the *valence band*.

An electron receiving energy can jump to a higher *conduction band* (or orbit), where it may roam through a substance.

The gap between these bands is called the *forbidden band*, and the amount of energy required to jump the gap is material-dependent.

Materials can be classified by their valence bands and band gaps thus:

Insulators are materials with (electron) full valence bands and high band-gap energies
   (>3 eV or > 4.8×10^{-16} J).
Conductors are materials with relatively empty valence bands and low band-gap
   energies.
Semiconductors are materials with relatively full valence bands and lower band-gap
   energies (< 3 eV or < 4.8×10^{-16} J).

Examples of semiconductors and their band-gap energies include copper indium diselenide (1.01 eV), silicon (1.11 eV), cadmium telluride (1.44 eV), cadmium sulphide (2.42 eV) and gallium arsenide (1.4 eV).

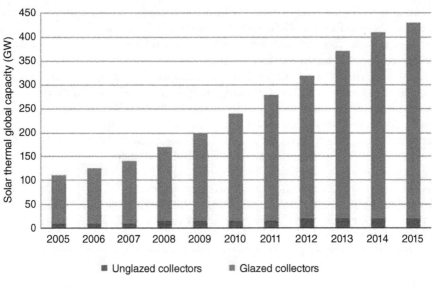

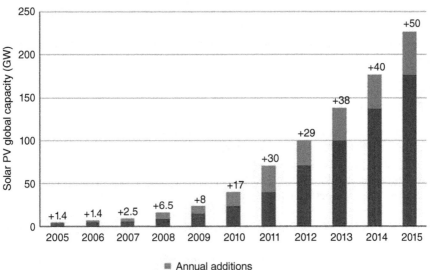

Figure 8.3 Solar thermal and solar PV global capacity 2005–2015. Data courtesy of REN21 Global status report 2016.

In PVs, photons of solar radiation are used to raise an electron from the valence band to the conduction band, making themselves available for current flow.

Photonic energy (W_{photon}, Joules) is calculated from:

$$W_{photon} = hf = hc/\lambda \tag{8.1}$$

where h is Planck's constant $= 6.625 \times 10^{-34}$ J s, f is the photon frequency (Hz), λ is the photon wavelength (m) and c is the speed of light $= 3 \times 10^8$ m/s in a vacuum.

Photons having less energy than the material band-gap energy will not raise any electrons, and photons with more energy will dislodge electrons but expend their surplus energy in heat.

Photovoltaic technology exploits this discovery to convert sunlight directly into DC electricity. Suitable power conditioning facilitates a conversion to AC electricity.

Four main types of PVs are currently available:

- *Mono-crystalline silicon* – These forms are blue/black and homogenous in appearance. They are sliced from a single silicon crystal. Cell conversion efficiencies are in the range 13–17%, with slightly lower module efficiencies of 12–15%.
  Typically, a single cell with dimensions of 100 mm × 100 mm can generate 3 amps at 0.5 volts, i.e. 1.5 watts in bright sun.
- *Polycrystalline silicon* – Polycrystalline forms are blue and multi-faceted in appearance. They are made from a silicon cast in a mould and consequently can be larger than monocrystalline forms. Cell conversion efficiencies are in the range 12–15%, again with slightly lower module efficiencies of 11–14%. They are cheaper than monocrystalline forms.
- *Thin film* – These are matt black, grey or brown in appearance. They are made by depositing an amorphous silicon, cadmium telluride or copper indium diselenide coating onto a glass front substrate. The substrate is laminated or polymer coated to provide climatic protection. Cell conversion efficiencies are in the range 5–10%, with lower module efficiencies of 4–7.5%. They are cheaper than polycrystalline forms.
- *Dye-sensitive polymers* – Their operation is unlike the PV forms described above, and it more closely resembles photosynthesis. They are manufactured from semiconductor titanium dioxide (TiO_2) and a conducting saline solution, printed onto film.

Although their conversion efficiency is much lower (< 5%), their construction materials are non-toxic and cheap to produce. Furthermore, they are relatively tilt and shading-insensitive and their efficiency increases with increasing ambient temperature.

For a single cell, the current is a maximum and the voltage zero under short circuit conditions. Under open circuit conditions, the current is zero and the voltage is a maximum. Between these two conditions, the current–voltage relationship varies, as shown in Figure 8.4.

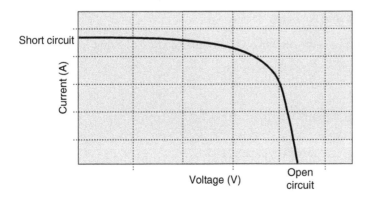

Figure 8.4 Typical current–voltage relationship for a PV cell.

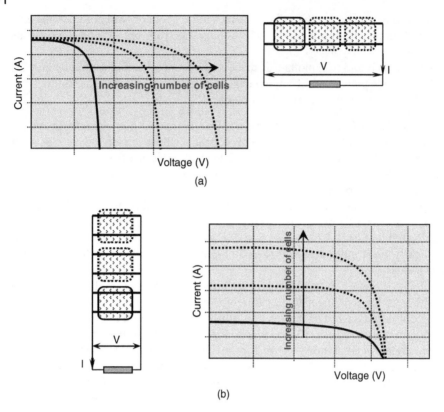

Figure 8.5 Effect of series/parallel arrangement on PV electrical characteristics. (a) PVs in series; (b) PVs in parallel.

To obtain useful voltages and currents, PVs are operated in a series/parallel, modular arrangement. For PVs in series, the voltages are additive at a given current and for PVs in parallel, the currents are additive at a single voltage, as shown in Figure 8.5.

The electrical DC power (W) generated by a single cell can be calculated from:

$$\text{Electrical power} = \text{Cell current} \times \text{Cell voltage}$$

The annual energy output (kWh/m^2) of a PV module can be estimated from:

$$W_{PV} = I_s \times (C_1/100) \times C_2 \times (C_3/100) \times C_4 \tag{8.2}$$

where:

I_s = Annual solar energy input (kWh/m^2)
C_1 = PV module efficiency (%)
C_2 = Shading and soiling de-rating – typically 0.95
C_3 = Electrical power system efficiency (%) – typically 85%
C_4 = Heat losses de-rating – typically 0.97.

8.3.2 Solar Thermal Power Plants

Many people associate solar electricity generation solely with photovoltaics and not with solar thermal power. Yet, large, commercial, concentrating solar thermal power plants have been generating electricity at a reasonable cost for a number of years.

Solar thermal power plants are designed to maximize the collection of thermal radiation from the sun by focusing and directing solar rays in order to concentrate thermal energy density.

Most techniques for generating electricity from heat need high temperatures to achieve reasonable efficiencies. The output temperatures of non-concentrating solar collectors are limited to below 200 °C. Concentrating systems are able to do much better. Due to their high costs, lenses are not usually used for large-scale power plants, and more cost-effective alternatives are used, including reflecting concentrators. The four most common reflecting arrangements are shown in Figure 8.6.

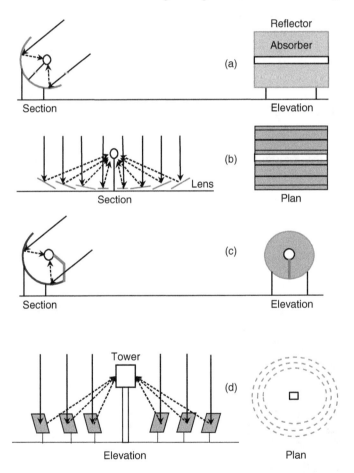

Figure 8.6 Solar thermal power generation plants. (a) Parabolic trough collector; (b) linear Fresnel collector; (c) central receiver system with dish collector; (d) central receiver system with distributed reflectors.

The reflector, which concentrates the sunlight to a focal line or focal point, has a parabolic shape; reflectors usually track the sun. In general terms, a distinction can be made between one-axis and two-axis tracking: one-axis tracking systems concentrate the sunlight onto an absorber tube in the focal line, while two-axis tracking systems do so onto a relatively small absorber surface near the focal point of the system.

8.4 Biomass Power Plants

The most common form of biomass is wood. For thousands of years, people have burned wood for heating and cooking. Wood was the main source of energy of the world until the mid-1800s and continues to be a major source of energy in much of the developing world.

The utilization of coal and the Industrial Revolution in the United Kingdom changed all that, as our energy consumption became more diverse and its uses spread to transport, machinery and leisure. The discovery of oil had even more impact on biomass, making it almost a redundant source of energy. This trend is now in reverse.

The current rate of global biomass power production is detailed in Figure 8.7.

Biomass is often considered a carbon-neutral fuel because the CO_2 released during combustion was originally extracted from atmospheric air during photosynthesis (although carbon emissions associated with its transport and processing are relevant). Photosynthesis uses solar energy and so biomass is essentially solar energy that has been converted into chemical energy and stored in organic matter (sugars, starches and cellulose).

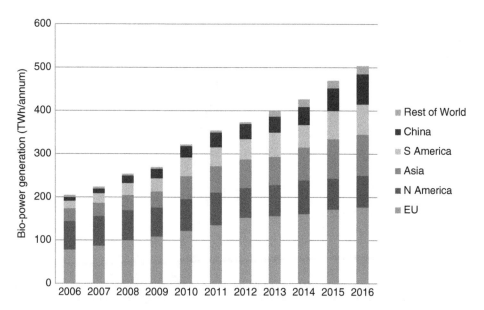

Figure 8.7 Global bio power generation by region, 2006–2016. Data courtesy of REN21 Global status report 2017.

Biomass can be derived from the following sources:

Farm waste – This category includes straw from cereals and pulses, stalks and seed coats of oil seeds, stalks and sticks of fibre crops, pulp and wastes of plantation crops, peelings, pulp and stalks of fruits and vegetables and other waste like sugarcane, rice husk, molasses, coconut shells etc.
Animal waste also constitutes a rich source of biomass that can be added to this category.
Public parks and forest waste – Logs, chips, bark and leaves collected from managed parks and forests in order to reduce fire hazard and general tidiness.
Industrial waste – This group includes paper waste, plastic and textile waste, gas, oil, paraffin, cotton seeds and fibres, bagasse etc. Plastic and rubber wastes, in particular, have good calorific values.
Municipal solid waste and sewage sludge – Municipal solid waste is a mixture, typically containing 40% paper and 20% organic matter (food), with the remainder comprising plastic, metal and glass. Sewage sludge contains organic matter and nutrients that can be utilized for the production of methane through anaerobic digestion.
Algae – Algae can be found in freshwater or in saline seawater conditions. Most species are photoautotrophic, converting solar energy into chemical forms through photosynthesis. Algae have received considerable interest as a potential feedstock for biofuel production because they can produce useful quantities of polysaccharides (sugars) and triacylglycerides (fats). These are the raw materials for producing bioethanol and biodiesel transport fuels.
Biomass for energy farming – Standalone plots of land used to grow trees and plants for the specific purpose of being used as fuel. Commonly used species include *miscanthus*, elephant grass and poplar trees.

The processes by which these sources are converted to fuel and utilized are detailed in Table 8.1.

8.4.1 Forestry, Agricultural and Municipal Biomass for Direct Combustion

Commercially available woody biomass fuels include pellets, logs and chips.

From a practical perspective, the four most important wood biomass fuel characteristics to consider are bulk density, ash content, moisture content and calorific value.

8.4.1.1 Bulk Density (kg/m^3)

This property describes the amount of space or volume required to accommodate 1 kilogram of fuel. Values vary; however, in general, wood pellets tend to have the greatest bulk density as they comprise mechanically compressed sawdust. Wood chips have the lowest bulk density as a consequence of their manufacture. Fuel storage plays an important part in a biomass energy system, and a fuel's bulk density, energy content and the time between deliveries will determine the fuel store size.

Typical bulk densities for woody biomass fuels are between 250 and 650 kg/m^3.

8.4.1.2 Moisture Content (% by Mass)

All woody fuels contain moisture or, more simply, water. When the fuel is burnt, some of its energy is used in converting the water to vapour, which is then usually lost up the

Table 8.1 Biomass conversion technology matrix.

Major biomass feedstock	Technology	Conversion process type	Energy or fuel produced
Wood agricultural waste Municipal solid waste Residential fuels	Direct combustion	Thermochemical	Electricity, heat
Wood agricultural waste Municipal solid waste	Gasification	Thermochemical	Low or medium-producer gas
Wood agricultural waste Municipal solid waste	Pyrolysis	Thermochemical	Synthetic fuel oil (bio crude) Charcoal
Animal manure Agricultural waste Landfill Wastewater	Anaerobic digestion	Biochemical (anaerobic)	Medium gas (methane)
Sugar or starch crops Wood waste Pulp sludge Grass straw	Ethanol production	Biochemical (aerobic)	Ethanol
Rapeseed soy beans Waste vegetable oil Animal fats	Biodiesel production	Chemical	Biodiesel
Wood agricultural waste Municipal solid waste	Methanol production	Thermochemical	Methanol

boiler flue or chimney. Keeping the moisture content in a biomass fuel to a minimum is therefore desirable and fuels are commonly either air- or kiln-dried before sale.

Typical moisture contents for woody biomass fuels are between 10 and 30%.

8.4.1.3 Ash Content (% by Mass)

This property describes the fuel mass percentage that will not burn, e.g. carbonate, oxide, hydroxide and nitrate compounds that remain after combustion. They will either remain in the boiler grate or, if entrained in the flue gases, be captured by the boiler grit arrester or filter. Ultimately, the ash content of a fuel will constitute a solid waste disposal problem. However, woody biomass ash makes a good fertilizer.

Typical ash contents for woody biomass fuels are in the range 0.5–2%.

8.4.1.4 Calorific Value (kJ/kg) and Combustion

Wood combustion processes are quite complex due to the nature of the fuel and its non uniformity. For dry wood (zero moisture content), the following simplified stoichiometric combustion equation (with oxygen) can be used to estimate

Table 8.2 Typical ultimate analysis of woody biomass.

Composition	% by weight
Hydrogen	6.0
Carbon	51.4
Nitrogen	0.4
Oxygen	41.3
Sulphur	0.0
Chlorine	0.0
Ash	0.9
Total	100.0

product/reactant quantities:

$$C_{3.17}H_{6.5}O_{2.71} + 4.44O_2 \rightarrow 4.17CO_2 + 3.25H_2O \tag{8.3}$$

The ultimate (i.e. elemental weight fraction) analysis of dry wood varies slightly from one species of plant to another, but a typical composition for a woody biomass is shown in Table 8.2.

In order to estimate calorific value, the reactions and energy releases of three elements, i.e. carbon, hydrogen and sulphur, with oxygen must be considered:

Combustion of carbon:

$$C + O_2 \rightarrow CO_2 + energy$$

Carbon has an energy content of 32 793 kJ/kg.
Combustion of hydrogen:

$$H_2 + \frac{1}{2}O_2 \rightarrow H_2O + energy$$

Hydrogen has an energy content of 142 920 kJ/kg.
Combustion of sulphur:

$$S + O_2 \rightarrow SO_2 + energy$$

Sulphur has an energy content of 9300 kJ/kg.

For example, the resulting calorific value of the fuel detailed in Table 8.2 can be determined by using the weighted mass fraction and component calorific values, thus:

Carbon contribution: $+ 32\ 793 \times 0.514 = 16\ 855.6$ kJ/kg

Hydrogen contribution: $+ 142\ 920 \times 0.06 = 8575.1$ kJ/kg

Sulphur contribution: $= 0$

During the combustion process, hydrogen reacts with oxygen to produce water. This will have a negative impact on the resultant calorific value.
Energy used in product water content

$$-2256.7 \times 0.06 = -135.4 \text{ kJ/kg}$$

Therefore

Net calorific value of fuel $= 16\,855.6 + 8575.1 - 135.4 = 25\,296$ kJ/kg

Typically, in practice, the combustion of dry wood releases about 10–20 MJ per kg.

The following empirical equation may be used to evaluate the higher heating value (HHV, kJ/kg) of biomass:

$$HHV = 0.3491C + 1.1783H - 0.1034O - 0.0211A + 0.1005S - 0.0151N \quad (8.4)$$

where C is the weight fraction of carbon; H of hydrogen; O of oxygen; A of ash; S of sulphur and N of nitrogen appearing in the ultimate analysis.

8.4.2 Anaerobic Digestion

Responsible gardeners do not dispose of non-woody plant and vegetation waste in their waste-disposal bins. They collect and store or 'compost' it in a container so that the products of the natural decomposition of the materials can be used as a fertilizer. Waste plant material used in this way is, quite often, just heaped in a corner of the garden. On the surface of the store, air-loving or aerobic bacteria break down the material into carbon dioxide, water and residue. However, deep inside the heap or store, there is little air and the bacteria carrying out the breakdown are said to be doing their work under *anaerobic* (air-absent) conditions.

(Incidentally, the anaerobic process also goes on in landfill sites, marshland and the guts of ruminants, for example, cows, sheep etc.)

When containers are used for compost storage, they are not usually air-tight. An anaerobic digester is a scaled-up storage container with the air deliberately excluded and seeded with bacteria that thrive in the absence of air.

As by-products of their metabolism, the bacteria in an anaerobic digester produce methane and carbon dioxide (*biogas*) and a solid/liquid fraction (*digestate*).

The composition of the biogas will vary depending on the biomass feedstock or 'substrate' and process conditions. Typically, a biogas might comprise approximately 65% methane and 35% carbon dioxide, with nitrogen, hydrogen, ammonia and hydrogen sulphide having trace ($< 1\%$) presences.

The energy content of biogas is in the range 17–25 MJ/m^3. This compares with 34 MJ/m^3 for natural gas.

The digestate is rich in nutrients (nitrates, phosphates, potassium) and can be used as a fertilizer.

Many different biomass waste sources or substrates can be used in anaerobic digestion, including animal slurries, silage, sewage sludge, food processing waste and household/kitchen waste. Woody wastes are, however, less suitable, as the bacteria find them difficult to metabolize. Important substrate properties include the proportion of substrate that is non-liquid, the proportion that is organic, i.e. composed of carbon and hydrogen compounds and will decompose to form a fuel, and the biogas yield. Energy output can be estimated from:

Annual biogas energy production (MJ/annum) = Biogas production (m^3/annum) × Biogas energy content (MJ/m^3)

where Biogas production (m^3/annum) = Mass of substrate (tonnes/annum) × [Volatile solid content (%)/100] × Biogas yield (m^3/tonne of volatile solid)

and Biogas energy content (MJ/annum) = 0.34 × [Biogas methane content (%)/100].

The biogas generation process takes place in three stages:

The addition of water (termed *hydrolysis*) to break down the waste matter (cellulose and proteins) into amino acids, fatty acids and glucose.

The conversion (*acidification*) of these products into organic acids, hydrogen and carbon dioxide by acid and acetate-forming bacteria.

The digestion of the organic acids and hydrogen by other bacteria, converting them to methane, carbon dioxide and water (this is termed *methanogenesis*).

Anaerobic bacteria at the centre of the process require specific conditions to thrive. Typical desirable conditions include:

The absence of oxygen.

A substrate with a minimum 50% moisture content.

An appropriate temperature depending on the strain of bacteria (supplementary heating is usually essential at northern European latitudes).

Sufficient retention time prior to production (25–40 days for temperatures of 30–40 °C).

A digester pH value of ~ 7.5.

Sufficient substrate surface area – may need chopping/grinding.

Good mixing in the digester to provide even conditions and avoid pressure build-up.

The absence of disinfectants and antibiotics in the substrate.

Distinct environments or zones develop within the reactor over time, as shown in Figure 8.8, resulting in liquid and solid outputs in addition to biogas.

Digester capacities in the range of 5–100 m³ are regarded as small scale, having typical annual substrate supplies in the region of 100–1000 tonnes.

Digester capacities over 100 m³ are classified as farm or large scale, having typical annual substrate supplies in the region of 1000–15 000 tonnes.

After cleaning and gas separation (i.e. CO_2, H_2S, O_2 etc.), biogas can have a range of uses, for example, electricity and heat production in a gas engine. Alternatively, if

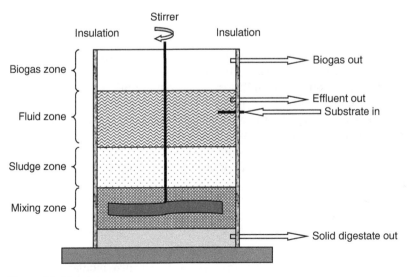

Figure 8.8 Digester zones.

its calorific value is boosted by the addition of propane, cleaned and pressurized, the resulting biogas (or *biomethane*) can be injected into the local gas distribution network.

8.4.3 Biofuels

Fuels like gasoline and diesel are based on the distillation and reforming of fossil fuels. The energy they release depends on the breaking of carbon and hydrogen molecular bonds that once constituted ancient life.

Current life on the planet is, of course, still based on the chemistry of carbon and hydrogen, with the important difference that these bonds are being continually re-made through naturally occurring biochemical cycles.

With a little knowledge of organic chemistry, contemporary carbon and hydrogen can be manipulated to produce renewable liquid fuels or *biofuels* (principally biodiesel and bioethanol) to be blended with non-renewable transport fuels.

8.4.3.1 Biodiesel

Around the turn of the 19th century, Rudolf Diesel demonstrated his new compression-ignition internal combustion engine in Paris. The fuel used by the engine was peanut oil. However, the fast-developing petrochemical industry soon produced a fossil-fuel-based alternative that was cheaper and more readily available.

Diesel's 'vegetable oil' fuel has merit and, in the 21st century, this idea is being revisited. In the last 100 years, our understanding of combustion and gas dynamics has improved greatly. It is now understood that reacting a basic vegetable oil with either ethanol (*grain alcohol*) or methanol (*wood alcohol*) results in a much-improved, diesel-like fuel, i.e. *biodiesel*. (A chemist or chemical engineer would describe the resulting fuel as an *ester* and the process as *transesterfication*.)

The process also results in a useful glycerol by-product.

$$\text{Vegetable oil} + \text{Methanol/Ethanol} \xrightarrow{\text{Catalyst}} \text{Biodiesel} + \text{Glycerol} + \text{Water}$$

Sources of vegetable oils (e.g. rapeseed, sunflower, palm, soya etc.) are commonly used in food processing and are readily and cheaply available. Even waste cooking oil and animal fats have potential as a feedstock for biodiesel.

How does the biodiesel compare with its fossil fuel competitor? For a given volume, the energy content of biodiesel at approximately 34 MJ/m^3 is typically 10–20% less than fossil-fuel diesel. However, compared to normal diesel, biodiesel produces fewer particulates, unburnt hydrocarbons and less carbon monoxide. Typically, 20% of biodiesel can be added to normal diesel as a blend in an unmodified engine.

Blends are usually signified by the use of B followed by the percentage of biodiesel in the blend, for example, B5, B10 and B20.

In Europe, most biodiesel uses rapeseed as a feedstock, with a typical biodiesel annual yield of 1.3 tonnes per hectare.

8.4.3.2 Bioethanol

Bioethanol is the renewable analogue to fossil-fuel gasoline. Unlike biodiesel it is commonly produced from glucose-rich plants like corn, sugar cane, beet or sorghum by *fermentation* in the presence of micro-organisms. In its simplest form:

$$\text{Simple sugars} \xrightarrow[\text{Micro-organisms}]{} \text{Ethanol} + CO_2$$

It is also possible to produce simple sugars and ethanol from starch-rich (grain and root) crops and cellulose (woody) plants. Cellulose feedstock, in particular, requires a little more processing (acid or enzyme *hydrolysis*); however, it is easier to grow and does not place the same demands on the soil as sugar-rich plants, making it suitable for marginal land, and therefore avoiding conflict with food crop production.

Micro-organisms are sensitive to their living environment and so the ethanol must be produced at low concentrations and distilled later.

How does the bioethanol compare with its fossil fuel competitor? For a given volume, the energy content of bioethanol at approximately 25 MJ/m^3 is typically 25–33% less than fossil-fuel gasoline. However, power loss is limited due to the better combustion efficiency of the biofuel.

Typically, 20% of bioethanol can be added to normal gasoline as a blend in an unmodified engine. With engine modification, use of up to 85–95% bioethanol is possible.

In Europe, typical bioethanol annual yields of 5 tonnes per hectare for sugar beet and approximately 2 tonnes per hectare for wheat are possible.

On the face of it, the use of liquid biofuels seems to provide a possible answer to carbon dioxide emissions associated with transport.

Some countries have a 'set-aside' policy for agricultural land, and bringing this land back into production for liquid biofuels would again appear logical.

Additionally, the large-scale take up of liquid biofuels could help reduce a country or region's dependence on imported oil.

However, the 'food vs. fuel' issue has to be considered, especially for countries that do not have a food surplus. To alleviate this dilemma, research and development of crops that are able to thrive on marginal land is required.

Furthermore, fossil-fuel-based fertilizer is often used to help produce the biofuel crop, and so the ratio of energy produced to fossil fuel consumed is relevant. The magnitude of this ratio varies with crop and land quality. This factor has given rise to 'second generation' biofuels based on agricultural and forestry waste.

8.4.4 Gasification and Pyrolysis of Biomass

Compared to gaseous fuels, the combustion of biomass is quite complex. For the combustion of methane with an oxidant, the process concerns the breaking of the gaseous reactant molecules and the formation of the gaseous product molecules, e.g. CO_2, H_2O. However, for the combustion of biomass, it is more accurate to describe the process as being a thermal decomposition of the fuel that boils off the fuel's water content and volatilizes its constituents to produce gases, condensable vapours and a char.

The process is sequential and temperature-dependent.
Some steps require little or no oxidant.
At 100–150 °C, the water content of a biomass fuel boils off.

At 200–500 °C and in an oxygen-starved environment, the wood is volatilized to form gases, a bio-char and a bio-tar. This low-temperature anaerobic process is termed

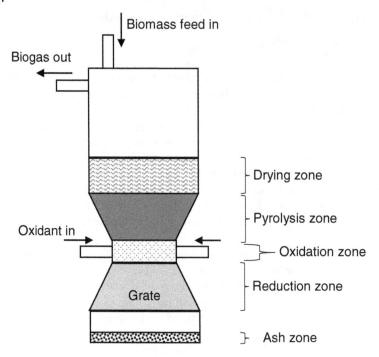

Figure 8.9 Biomass gasifier.

pyrolysis. (Roasting between 200–320 °C is termed *torrefaction.* This produces a superior solid biomass fuel in terms of transport, storage and suitability for further processing.)

At 500–800 °C, the gases would normally combust; however, if the biomass is denied an oxidant at 650–800 °C, the bio-char is further converted into complex flammable gases ('producer gas'), and above 800 °C, any remaining tar residues are broken down into simpler components (H_2, CO) which can be reformed into complex fuels. This high-temperature reduction process is termed *gasification.*

If the combustion phase is retarded by supplying insufficient oxidizer, it is possible to control the process to maximize the output of condensable vapours and flammable gases (CH_4, H_2 and CO). The control of oxidant (whether oxygen, air or steam), heat supply rate and residence time is at the heart of the pyrolysis and gasification of biomass. The processes are not new and the chemistry for other solid fuels such as coal has been practised for around 100 years. A typical biomass gasifier is illustrated in Figure 8.9.

8.5 Geothermal Power Plants

The word 'geothermal' is rooted in two Greek words, 'geo' (earth) and 'therme' (heat). The inner earth's heat is the result of planet formation from dust and gases that coalesced billions of years ago. Since the radioactive decomposition of elements in

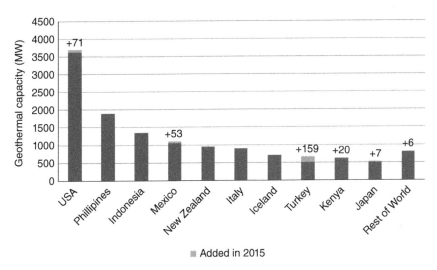

Figure 8.10 Geothermal power capacity for top ten countries in 2015. Data courtesy of REN21 Global status report 2016.

rocks continuously regenerates this heat, *deep* geothermal energy is plentiful and can be considered a renewable energy resource.

An indication of global geothermal generating capacity is shown in Figure 8.10.

Deep geothermal energy sources can be used for the central heating of buildings and for generating electricity. Geothermal gradients are typically 20–40 °C/km.

The largest geothermal system used for heating is located in Reykjavik, Iceland, and 89% of households in Iceland are heated in this way. Geothermal energy is also widely exploited in some areas of New Zealand, Japan, Italy, the Philippines and parts of the USA.

The most common way of capturing the energy from geothermal sources is to tap into naturally occurring 'hydrothermal convection' rock strata systems via wells. Steam at high temperature (up to 150 °C) and pressure from the extraction well is transported to the power plant to drive the steam turbine; upon its exit, it is taken back to an injection well in order to be reheated.

There are three basic designs for geothermal power plants:

Dry steam plant: Used with a geothermal source having a high dryness fraction. Steam is extracted directly from the rock store and transferred to the turbine. The steam exiting the turbine is condensed for return to the strata.

Flash steam plant: Used with a geothermal hot water source. Water is depressurized or 'flashed' into steam and then used to drive the turbine. The 'wet' brine fraction from the flash vessel is transferred to the turbine exit.

Binary geothermal plant: Hot water is passed through a heat exchanger, where it heats a second liquid – such as isobutane – in a closed loop. Isobutane boils at a lower temperature than water and is easily converted into vapour to run the turbine.

The three varieties of power plant are shown in Figure 8.11.

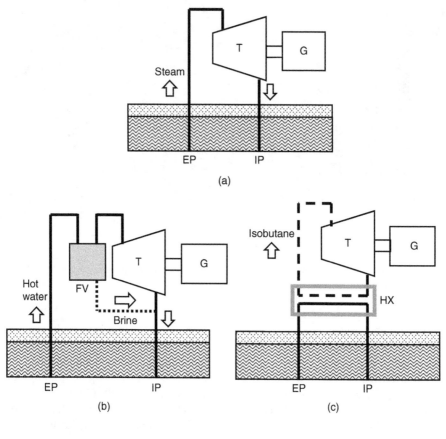

Figure 8.11 Geothermal steam power plant variations. (a) Dry steam; (b) flash steam plant; (c) binary plant.

8.6 Wind Energy

Wind energy was utilized for river transport along the Nile as early as 5000 BCE. Evidence of non-transport, wind power applications can be documented as far back as the first century CE in China and the Middle East, where windmills were used to pump water and grind grain.

The UK is the windiest country in Europe and has potentially the largest offshore wind energy resource in the world. It is estimated that there is enough energy available from this form of renewable energy to power the country almost three times over.

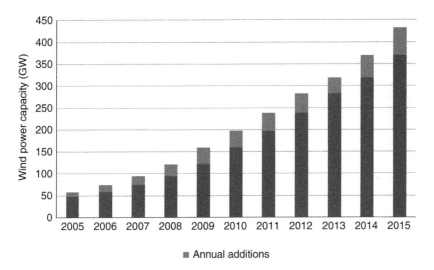

Figure 8.12 Global wind power 2005–2015. Data courtesy of REN21 Global status report 2016.

A wind farm is a collection of wind turbines combined to produce a prescribed output to satisfy the demand required. There are two categories of wind farm: onshore and offshore.

Wind turbine farms are not completely without problems. Visual impact causes concern in some quarters. Noise pollution can be an issue onshore in areas very close to installations. Offshore farms may interfere with the marine habitat. However, the technology is carbon-free and leaves a very small installation footprint.

Globally, 433 Gigawatts of wind-generating capacity was added to the grid between 2005 and 2015; see Figure 8.12.

8.6.1 Theory of Wind Energy

A wind turbine extracts power from the wind by slowing it down. At stand still, the rotor obviously produces no power; at very high rotational speeds, the air is more or less blocked by the rotor, and again no power is produced.

The power produced ($\dot{W}_w$) by the wind turbine is determined from the net kinetic energy change across it (i.e. from the initial air velocity of $\bar{V}_1$ at inlet to a turbine exit air velocity of $\bar{V}_2$, see Figure 8.13) and is described by:

$$\dot{W}_w = \frac{1}{2}\dot{m}(\bar{V}_1^2 - \bar{V}_2^2) \tag{8.5}$$

The mass flow rate of wind is given by the continuity equation as the product of density, area swept by the turbine rotor and the approach air velocity:

$$\dot{m} = \rho A \bar{V} \tag{8.6}$$

Hence, the power becomes:

$$\dot{W}_w = \frac{1}{2}\rho A \bar{V}_{ave}(\bar{V}_1^2 - \bar{V}_2^2) \tag{8.7}$$

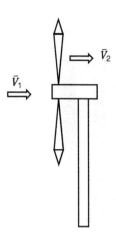

Figure 8.13 Ideal wind energy theory.

where $\bar{V}_{ave}$ is the average speed between inlet and outlet:

$$\bar{V}_{ave} = \frac{1}{2}(\bar{V}_1 + \bar{V}_2) \tag{8.8}$$

Substituting, the power becomes:

$$\dot{W}_w = \frac{1}{4}\rho A \bar{V}_1^3([1 - (\bar{V}_2/\bar{V}_1)^3 - (\bar{V}_2/\bar{V}_1)^2 + (\bar{V}_2/\bar{V}_1)]) \tag{8.9}$$

To find the maximum power extracted by the rotor, differentiate Equation (8.9) with respect to $\bar{V}_2$ and equate it to zero:

$$\frac{d\dot{W}_w}{d\bar{V}_2} = \frac{1}{4}\rho A(-3\bar{V}_2^2 - 2\bar{V}_1\bar{V}_2 + \bar{V}_1^2) = 0 \tag{8.10}$$

Since the area of the rotor (A) and the density of the air (ρ) cannot be zero, the expression in the bracket has to be zero. Hence, factorizing the quadratic equation:

$$(3\bar{V}_2 - \bar{V}_1)(\bar{V}_2 + \bar{V}_1) = 0$$

Since $\bar{V}_2 = -\bar{V}_1$ is unrealistic in this situation, there is only one solution:

$$\bar{V}_2 = \frac{1}{3}\bar{V}_1 \tag{8.11}$$

Substitution of Equation (8.11) into Equation (8.8) results in:

$$\dot{W}_w = (0.5925) \times \frac{1}{2}\rho A \bar{V}_1^3 \tag{8.12}$$

Equation (8.12) clearly shows that the power:

Is proportional to the density (ρ) of the air, which varies slightly with altitude and temperature.

Is proportional to the area (A) swept by the blades and thus to the square of the radius (R) of the rotor.

Varies with the cube of the wind speed ($\bar{V}^3$) This means that the power increases eightfold if the wind speed is doubled. Hence, one has to pay particular attention with respect to site selection.

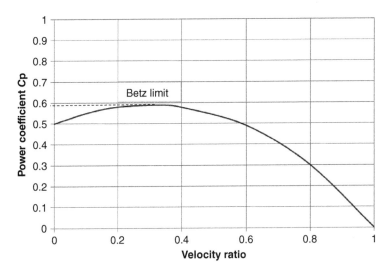

Figure 8.14 Variation in power coefficient with velocity ratio.

The value in brackets in Equation (8.12) (0.5925) is termed the *Betz limit* and indicates the theoretical maximum fraction of the power in the wind that could be extracted by an ideal wind turbine. Maximum efficiency depends on optimal wind velocity ratio conditions.

Away from optimal conditions, the theoretical output of the wind rotor is adjusted by the application of a power coefficient, C_p. The variation in power coefficient with wind velocity ratio is shown in Figure 8.14.

8.6.1.1 Actual Power Output of the Turbine

Equation (8.12) does not take into account the mechanical and electrical losses incurred in the production of electricity from a wind turbine. Allowing for non-optimal aerodynamic conditions and gearbox/electrical generator efficiencies (η_{gb} and η_{gen}), the following can be used to predict the shaft power of a wind rotor:

$$\dot{W}_e = \frac{1}{2}C_p\rho A \bar{V}^3 \eta_{gb}\eta_{gen} \tag{8.13}$$

The effect of these factors can be examined by considering the world's largest wind turbine generator which has a rotor blade diameter of 126 metres (rotor swept area 12 470 m^2) and is located offshore at sea level where the air density is approximately 1.2 kg/m^3.

At its rated wind speed of 14 m/s, the theoretical wind power produced is given by Equation (8.12) and is approximately 12 MW. However, the turbine is rated at 5 MW.

The difference in values indicates the influence of the power coefficient C_p and inefficiencies in the electrical and mechanical power transmission.

Care must be taken when selecting a turbine and assessing the likely energy generation, as the 'rated power' is only available when the wind turbine speed reaches a certain level, known as the *rated speed*. Wind turbine speed performance is split into four different stages:

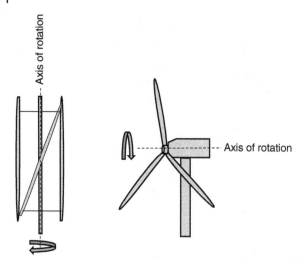

Figure 8.15 Wind turbine
types – vertical and horizontal axis.

1. *Start-up speed* – The speed at which the rotor and blade assembly begin to rotate.
2. *Cut-in speed* – The minimum wind speed at which the wind turbine will generate *any* usable power. This wind speed is typically between 3 and 4.5 m/s for most turbines.
3. *Rated speed* – The minimum wind speed at which the wind turbine will generate its *designated rated* power.
4. *Cut-out speed* – The wind speed at which the turbine is intentionally shut down. This is a safety feature to protect the wind turbine from damage. It is facilitated using an automatic brake or 'spoilers', which are drag flaps mounted on the blades or the hub, activated by excessive rotor revolutions. Alternatively, the entire machine may be turned sideways to the wind stream at high wind speeds.

8.6.2   Wind Turbine Types and Components

There are two types of wind turbine, distinguished by the axis of rotation of the rotor shafts:

1. Horizontal-axis wind turbines, also known as HAWT types, have a horizontal rotor shaft and an electrical generator, both of which are located at the top of a tower. The gearbox/generator housing is termed the *nacelle*.
2. Vertical-axis wind turbines (VAWT types) are designed with a vertical rotor shaft and a generator and gearbox placed at the base of the turbine. The turbine has a uniquely shaped rotor blade that is orientated to harvest the power of the wind no matter in which direction it is blowing.

Examples are shown in Figure 8.15.

8.7   Hydropower

Hydropower is power derived from the force of moving water. Hydropower is a versatile, flexible technology that, at its smallest, can power a single home and, at its largest,

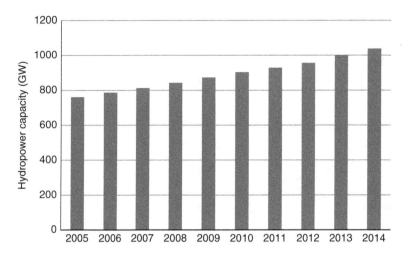

Figure 8.16 World capacity of hydropower. Data courtesy of EIA, international energy statistics.

can supply industry and the public with renewable electricity on a national and even inter-regional scale.

Hydropower is pollution-free, has a fast response time and can provide essential back-up power during major electricity outages or disruptions. Hydropower plants also provide a potential flood-control mechanism, extracting excess water from rivers. Finally, as part of a multipurpose scheme, impoundment hydropower creates reservoirs that offer a variety of recreational opportunities, notably fishing and water sports.

Hydropower is the largest source of renewable energy generation worldwide. In 2015, the total global capacity was 1064 GW, with a generation of 3940 TWh (REN21, 2016).

Global installed hydropower capacity was in the region of 1000 GW in 2010 and the trend remains upwards. The world leaders in hydropower are China, Brazil, Canada, the United States and Russia. Together these countries account for 52% of total installed capacity. Norway's electricity-generation system is almost 100% hydro, and there are a number of countries in Africa that produce close to 100% of their grid-based electricity from hydro (see Figure 8.16).

8.7.1 Types of Hydraulic Power Plant

There are two types of hydropower plant in use: run-of-river schemes and storage systems.

8.7.1.1 Run-of-river Hydropower

Run-of river hydropower schemes (see Figure 8.17) require a bifurcation or separation of a proportion of the flow via a weir *impoundment* system (1).

The take-off canal (2) is commonly called the *headrace* or *leat*.

At the end of the headrace is a settling tank (3), often called the *forebay*.

The forebay/headrace may contain *spillways* (4) to return excess flow to the river. From the forebay, the water enters a pipe (5), commonly called the *penstock*, to commence the final leg of its journey down to the hydro-turbine. If a site has difficult topography or

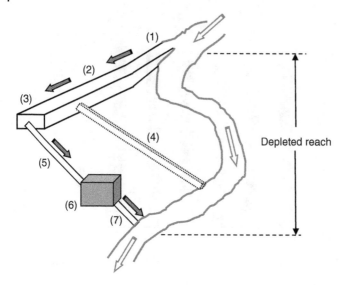

Figure 8.17 Run-of-river system characteristics.

is environmentally sensitive, the canal is sometimes omitted and an extended penstock carries water from the river bifurcation direct to the powerhouse.

The hydro-turbine and electrical generator are sited in the *powerhouse* (6).

After passage through the turbine, the water is returned to the river or stream via a calming channel (7), known as the *tailrace*.

In a *barrage* scheme arrangement, the headrace and penstock are dispensed with and the powerhouse is effectively *in-line* with the river or stream.

Whether considering an impoundment or a barrage scheme, the fraction of flow diverted for energy generation must not leave the depleted region (*reach*) of the river ecologically damaged, and consequently, extensive local hydrological and biological studies are carried out prior to scheme sizing.

8.7.1.2   Storage Hydropower

Typically, these are large systems that use a dam to store water in a reservoir (see Figure 8.18). Electricity is produced by releasing water from the reservoir through a turbine, which activates a generator.

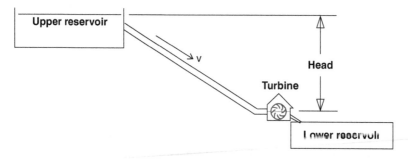

Figure 8.18 Typical storage hydropower plant.

Storage hydropower provides base load as well as the ability to be shut -down and started up at short notice according to the demands of the system (peak load). It can offer enough storage capacity to operate independently of the natural hydrological inflow for many weeks or even months.

8.7.2 Estimation of Hydropower

In a *storage* hydropower system, the amount of power produced from water is proportional to the potential energy available, due to the difference of *head* between the surface levels of the upper and lower reservoirs and the mass flow rate of water going through the turbine.

For a reservoir head difference of h (m), the general equation for fluid power $\dot{W}$ (W) is:

$$\dot{W}_{fluid} = \rho g h \dot{V} \tag{8.14}$$

However, the actual power produced by the turbine unit is less than is indicated by Equation (8.14) because of frictional losses (Δh_f, m) in the transmission pipework and the turbine's efficiency (η_t, %). If the efficiency of the electrical generator is (η_{gen}, %), the power output ($\dot{W}_{elec}$, W) of the system is:

$$\dot{W}_{elec} = \rho g (h - \Delta h_f) \dot{V} \eta_t \eta_{gen} \tag{8.15}$$

Run-of-river schemes are also evaluated in terms of flow rate (m³/s) and head difference or river fall (m).

8.7.3 Types of Hydraulic Turbine

Traditionally, there are three different types of turbine used in hydropower generation: the Pelton wheel, the Francis turbine and the Kaplan turbine, each of which has different characteristics to suit the available head (m) and discharge or flow rate (m³/s) (see Figure 8.19).

For low-head, run-of-river schemes, technology with a long history is being employed – the Archimedean screw. This technology was first used over 2000 years ago as a pump for lifting water from a river to its bank. However, in power-generation schemes, a portion of the flow is diverted and the head variation or fall along the river (perhaps via a weir) is used to drive the screw. The screw shaft is connected to a generator, as shown in Figure 8.20.

8.8 Wave and Tidal (or Marine) Power

Large-scale water waves on the earth's surface are largely caused by the gravitational pull of the moon (the tides) and underwater earthquakes ('tsunami'). Smaller-scale water surface disturbances or waves are generally caused by the *wind* blowing across a water surface. Wind-generated waves have the potential for energy extraction because water has been 'lifted up' from its undisturbed level and given a circular motion. The motion, i.e. the kinetic energy available, extends below the water surface but decreases with increasing depth.

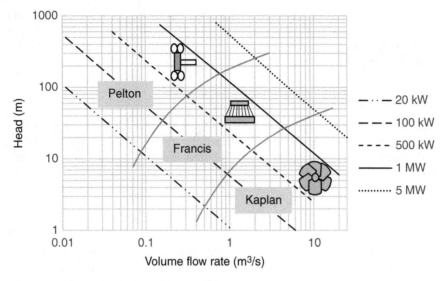

Figure 8.19 Typical hydraulic turbine specifications.

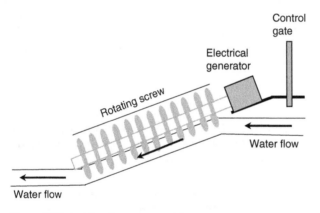

Figure 8.20 Archimedean screw turbine.

8.8.1 Characteristics of Waves

The motions of real water waves are complex. Simple analyses assume they have a regular repeating pattern of crests and troughs, as shown in Figure 8.21.

The distance between any two repeating points on the wave is called the *wavelength* (metres). The time taken for a wave to travel a distance of 1 wavelength, that is to complete a cycle, is called the *period* (T, seconds). Typical values of ocean wave periods are in the range 1–30 seconds. Storm and earthquake waves can have periods ranging from minutes to hours.

This factor is important, as the power (watts) available in a water wave increases directly with period, so, for example, a wave with a period of 10 seconds will, potentially, have two times more power than a wave with a period of 5 seconds.

The maximum vertical disturbance distance from wave crest to wave trough is called the *wave height* (H, metres) and depends on the prevailing wind regime.

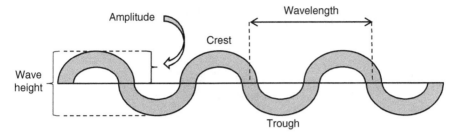

Figure 8.21 Wave variables.

The term *amplitude* (*a*, metres) is used to describe ½ the wave height.

This factor is important, as the power (watts) available in a water wave increases with the square of the amplitude, so, for example, a wave with an amplitude of 2 m will, potentially, have four times more power than a wave with an amplitude of 1 m.

8.8.2 Estimation of Wave Energy

The energy per metre of wave crest and unit of surface is given as:

$$W = \frac{\rho g a^2}{2} \tag{8.16}$$

where ρ (kg/m³) is the density of water.

It can be shown that the wave power available is given by:

$$\dot{W}_{wave} = \frac{\rho g^2}{32\pi}H^2 T \tag{8.17}$$

8.8.3 Types of Wave Power Device

There are several varieties of wave energy converter (WEC).

If deep water is available close to the shoreline, waves may be directed up a focusing ramp arrangement over a sea wall into a collection tank or reservoir in a so-called *wave capture* system. At the base of the reservoir is a low-head water turbine through which water is returned to the sea.

Essentially, this arrangement is recreating a typical hydroelectric scheme with seawater (see Figure 8.22).

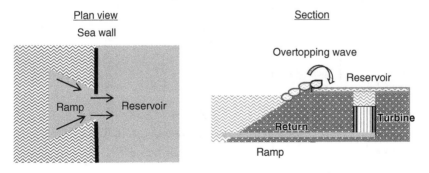

Figure 8.22 Wave power device – wave capture absorber.

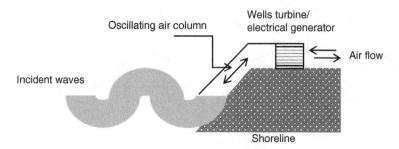

Figure 8.23 Wave power device – oscillating-column absorber.

In *oscillating-column* devices, the up and down motion of the water wave is used to create a mirrored motion in a volume of enclosed air (see Figure 8.23). The air can only exit its enclosure by passing through an air turbine connected to an electrical generator. The air direction will, of course, change as the water wave rises and falls, but this is overcome by use of a machine called a *Wells turbine* whose rotational direction is insensitive to the flow direction of the air. Shoreline and floating varieties of this kind of device have been demonstrated to have potential for significant energy generation.

Point absorber wave energy devices usually employ a float to utilize the reciprocating aspect of a wave. Energy conversion and take-off can take a range of forms, as shown in Figure 8.24. For example, the resulting motion can be used to drive a simple hydraulic pump to transfer a fluid to a reservoir and generator set.

Alternatively, the reciprocating motion can be used directly to drive a permanent magnet linear generator (a set of coils and magnet stack) mounted on the seabed.

These devices have the advantage of being insensitive to wave direction (see Figure 8.24).

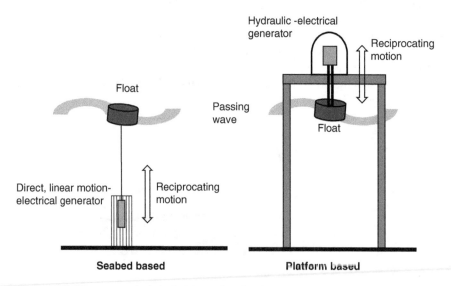

Figure 8.24 Wave power device – floating point absorber.

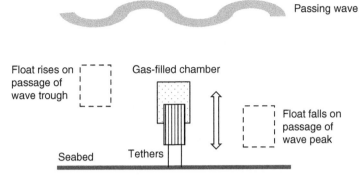

Figure 8.25 Wave power device – static-pressure point absorber.

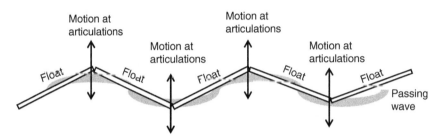

Figure 8.26 Wave power device – wave-profile-following device.

Point absorbers may also be completely submerged static-pressure devices. In this case, as a wave with its peaks and troughs passes above the device, a gas-filled chamber, tethered to the seabed, will experience changes in pressure (due to the height of water above it), which cause it to move up and down relative to a fixed power take-off point (see Figure 8.25). Again, the motion can be used to drive a pumping system or to generate electricity directly. This arrangement has the advantage of not exposing the system to the most damaging aspects of oceanic existence in the wave zone.

Floating and semi-submerged wave profile or line absorber devices typically rock back and forth and side to side in response to incident waves, i.e. they ride the waves instead of responding solely to the vertical motion. They often comprise replicated, articulated units (see Figure 8.26). The motion experienced at the universal joints is transferred to hydraulic rams that create pressure in a hydraulic system connected to electrical generators.

Wave surge devices comprise hinged-flap structures that oscillate back and forth in a horizontal plane (like a pendulum) with the passage of a wave (see Figure 8.27). The motion is converted into fluid pressure by, for example, a piston arrangement attached to the flap. The piston motion can be used to pump a fluid to a reservoir and electrical generator set.

8.8.4 Tidal Power

The moon and earth are locked together, with a mutual force of gravitational attraction keeping them wedded. The effect is complicated by the planet's oceans, since the liquid

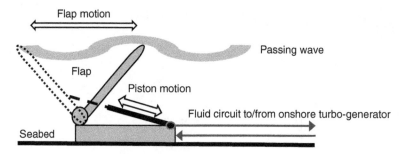

Figure 8.27 Wave power device – wave surge system.

and solid parts of the planet feel the pull of the moon to a different extent. The oceans feel it more and, in fact, bulge out towards it in its orbit (and a second bulge is generated on the other side of the planet). This effect is further enhanced, but to a lesser extent, by the proximity of the sun.

As the earth rotates under the bulges, the oceans ebb and flow at the coastline. These are the *tides*. The difference in height between the highest and lowest level of the tide is termed the *tidal range*. Out in the deep ocean, the height of the swelling is of the order of 1 m; however, at the ocean's edge, the reducing depth and shape of the coastline can modify the phenomenon, producing much larger tidal ranges. The greatest tidal range in the world is found in the Bay of Fundy (~ 16 m) in Canada. The second highest tidal range (~ 15 m) is found in the Severn estuary in the UK.

The time between bulges crossing a point on the planet is approximately 12 hours 25 minutes and is entirely predictable, unlike other forms of renewable energy like wind and solar, making this source very attractive from a power supply management point of view.

This effect can be harvested in two different ways.

8.8.4.1 Tidal Barrage Energy

In tidal barrage schemes, a sea dam is constructed between two adjacent points on a coastline, for example, an estuary, forming a man-made basin. The basin water volume and difference between water levels across the dam can be controlled via sluice gates. The dam also contains a series of turbo-generators, making this arrangement, in essence, very similar to a hydroelectric system, in that the energy is produced as a result of a difference in head (h) across the turbine. At its simplest, the greater the tidal range, the greater the potential for energy production. With the utilization of turbines that are insensitive to flow direction, the sluice/turbine arrangement has a good degree of flexibility for ebb/flow generation and/or making the basin available for storage considerations.

Depending on maritime traffic, barrage schemes may require a navigation lock. However, barrages can also double as bridges, providing new road links across the chosen estuaries.

The general characteristics of a tidal barrage scheme are shown in Figure 8.28.

The average fluid power generated by a scheme can be estimated from Equation (8.14), i.e. $\dot{W} = \rho g h \dot{V}$.

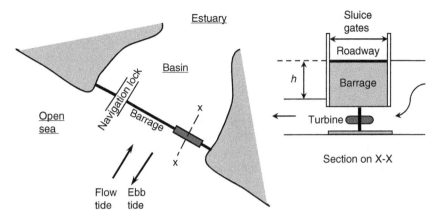

Figure 8.28 General characteristics of a tidal barrage scheme.

In the case of a tidal scheme, the basin volume is the product of basin area (A) and the tidal range (h). The tidal flow rate will be the basin volume divided by the time between tides, i.e. 4.47×10^4 seconds.

Finally, the power is calculated using the tidal range mid-point ($h/2$) thus:

$$\dot{W} = \rho g \times \frac{Ah}{4.47 \times 10^4} \times \frac{h}{2}$$
$$= \frac{\rho g A h^2}{8.94 \times 10^4} \tag{8.18}$$

8.8.4.2 Tidal Stream Energy

These systems take advantage of the high-speed ocean flows that can occur between islands or between an island and a coastline. The energy-collection devices often sit on the seabed and resemble submerged wind turbines.

8.9 Thermoelectric Energy

Energy harvesting is a term commonly applied to the scavenging of a previously untapped *small-scale* ambient source of energy. Sources of energy for harvesting include light, thermal gradients, vibration and radio frequency radiation. The scale of collection is typically from microwatts to milliwatts. There are many potential demands for this power level, with applications including condition monitoring in construction, health and general manufacture. Energy scavenged from ambient sources may be able to recharge or even eliminate the requirement for a battery in some mobile applications.

Research in this area is being conducted in parallel with energy storage devices (e.g. rechargeable batteries or supercapacitors) and efficient power management circuits to provide a stable power source.

There are many energy-harvesting conversion technologies available. Here, the direct conversion of thermal energy to electricity and thermoelectric generators will be considered.

8.9.1 Direct Thermal Energy to Electrical Energy Conversion

The discovery of direct thermal to electrical energy conversion is not recent. In early-1830s Berlin, Thomas Johann Seebeck heated the junction of a circuit comprising two different metals. He detected a magnetic field close to the circuit but failed to make the link with the presence of a voltage in the circuit. Nevertheless, his work is celebrated by the term *Seebeck effect.*

Consider the open circuit in Figure 8.29 comprising two *different* material electrical conductors forming a junction. With a suitable electron intensity difference between the two materials at the heated junction, the application of a little heat input is sufficient to allow electron flow, realizing a potential difference detectable across the circuit.

The magnitude of the voltage is proportional to the temperature of the junction and the chosen materials.

The slope of the voltage–junction temperature relationship is termed the *Seebeck coefficient* (α_{ab}, V/K) for the combination of materials a and b.

Mathematically:

$$\alpha_{ab} \equiv \lim_{\Delta T \to 0} \left(\frac{\Delta V}{\Delta T} \right) \text{ or } V = \int_{T_1}^{T_2} \alpha_{ab} dT \tag{8.19}$$

where $\alpha_{ab} = \alpha_a - \alpha_b$.

The absolute Seebeck coefficient of an individual material (α_i) can be determined by connecting it in a circuit with a superconductor. A comparison of some material Seebeck coefficients at 100 °C is shown in Table 8.3.

Notice that some material coefficients are quoted as positive and some negative. This convention is determined by the resulting flow direction in the circuit.

In 1834, French physicist Jean Peltier applied an electric current to a circuit comprising two dissimilar metals. He discovered that one of the circuit junctions increased in temperature while the other was cooled. He also found that the temperature change depended on the magnitude of the current and that if the current flow was reversed, the junction conditions reversed.

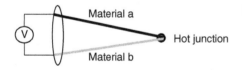

Figure 8.29 Seebeck effect.

Material a

Hot junction

Material b

Table 8.3 Seebeck effects for some common materials.

Material	Seebeck coefficient (α, V/K)
Silicon	-455×10^{-6}
Constantan	-47×10^{-6}
Platinum	-5.2×10^{-6}
Aluminium	-0.2×10^{-6}
Copper	$+3.5 \times 10^{-6}$
Iron	$+13.6 \times 10^{-6}$
Germanium	$+375 \times 10^{-6}$

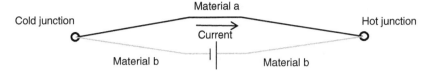

Figure 8.30 Peltier circuit.

Consider the circuit in Figure 8.30 comprising two different material electrical conductors forming two junctions and a voltage source.

With a current flowing in the circuit, it is found that heat is absorbed at one of the junctions (cold junction) and released at the other (hot junction).

The magnitude of the effect is expressed by the *Peltier coefficient* (π_{ab}, volts), which is a function of current (I) and is defined thus:

$$\pi_{ab} = \left(\frac{Q}{I} \right) \tag{8.20}$$

where Q is the heating (or cooling) effect.

This is essentially a phenomenon coupled with the Seebeck effect and the two are linked by:

$$\pi_{ab} = T\alpha_{ab} \tag{8.21}$$

where T is the absolute junction temperature in Kelvin.

8.9.2 Thermoelectric Generators (TEGs)

In energy-harvesting terms, a Peltier device consists of a circuit comprising two dissimilar materials, two junctions, a heat source, a heat sink and a connected external load. A current can be generated in the bi-material circuit when one junction is heated and the other cooled.

Potentially (p and n type) semiconductors make better thermoelectric generators than metals because of their higher Seebeck coefficients.

A schematic of a semiconductor, single junction thermoelectric generator is shown in Figure 8.31.

For useful power production, such units are usually connected in a series/parallel arrangement where a number of p-type and n-type semiconductor materials are sandwiched between electrically insulated ceramic plates joined together by interconnected copper pads.

Figure 8.31 Thermoelectric generator.

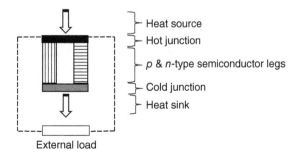

The power generated in the thermoelectric circuit ($\dot{W}_{TE}$, Watts) is given by:

$$\dot{W}_{TE} = \pi_{ab}I = \alpha_{ab}I\Delta T \tag{8.22}$$

From Ohm's Law and the Seebeck voltage–temperature relationship:

$$I = \frac{V}{R_{TE}} = \frac{\alpha\Delta T}{R_{TE}} \tag{8.23}$$

The resistance of the generator is given by:

$$R_{TE} = \frac{\rho L}{A} \tag{8.24}$$

where A (m^2) is the cross-sectional area of the TEG circuit, L (m) its length and ρ its resistivity (Ω/m).

Substituting into the above, the total power generated by a thermoelectric device is given by:

$$\dot{W}_{TE} = \Delta T^2 \frac{\alpha^2 A}{\rho L} \tag{8.25}$$

The thermoelectric generator will be connected in series with the external load. The power expended at an external load will depend on its resistance and resulting current.

For the device described, the efficiency will be the power available across a connected load divided by the power input to the hot junction or heat source.

$$\eta = \dot{W}_{TE}/\dot{Q}_{hot} \tag{8.26}$$

In general, efficiency values for thermoelectric generators tend to be low, with values below 10% being commonly reported.

However, they have no moving parts, are compact and can use almost any source of thermal energy as the hot source.

8.10 Fuel Cells

Chemical reactions are, to a great extent, concerned with the rearrangement of electrons between chemical species. In a combustion reaction between hydrogen and oxygen, this results in a great deal of thermal energy being released that can be used to generate electricity by heating up a working fluid and using it to drive some kind of prime mover like a steam or gas turbine at low efficiency.

In a fuel cell, the electron exchange between the hydrogen and oxygen contributes directly to the generation of an electrical current without the need for any intermediate stages.

At a basic level, the components of a fuel cell are similar to those of a battery: anode, cathode, electrolyte and external circuit.

However, unlike a battery, in a fuel cell, complete discharge cannot occur because the cell is constantly supplied with electron donators and acceptors in the form of hydrogen and oxygen. On combination, the reactions produce an electrical current in an external circuit, water and some waste heat. The generic components of a fuel cell are shown in Figure 8.32.

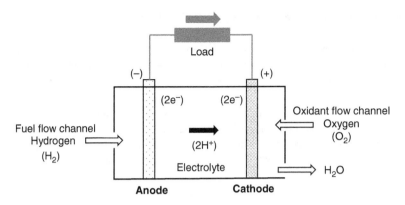

Figure 8.32 Generic fuel cell.

8.10.1 Principles of Simple Fuel Cell Operation

At the anode, hydrogen undergoes a catalytic reaction, splitting into ions ($2H^+$) and electrons ($2e^-$).

Ions traverse the electrolyte; electrons flow through the external circuit.

At the cathode, hydrogen ions, oxygen and electrons from the external circuit form water in a catalyst-assisted reaction.

A simple arrangement such as this might generate up to 1 volt DC.

Practical fuel cells comprise repeating units of the above in series, forming a *stack* to produce a more useful level of power. The number of repeating units determines the voltage and the ionic transfer (or transverse) area determines the current.

Electronic conditioning is required to produce AC power with grid-compatible parameters.

With hydrogen as the fuel, the technology is carbon-free at the point of generation. Questions remain over the source of hydrogen for use in such cells. Current potential sources include separation from hydrocarbons (e.g. methane can be reformed into H_2 and CO_2) or the splitting of water. Both are potentially energy-intensive.

8.10.2 Fuel Cell Efficiency

Care must be taken when quoting the efficiency of a fuel cell. In thermodynamic terms, the *reversible cell reaction efficiency* ($\eta_{reaction}$) is defined by considerations of molar gas flows ($\dot{n}$, moles), Gibbs energies ($\bar{g}$, kJ/mol) and enthalpies ($\bar{h}$, kJ/mol) at a given temperature and pressure:

$$\eta_{reaction} = \frac{\sum\limits_{in} \dot{n}_i \bar{g}_i - \sum\limits_{out} \dot{n}_i \bar{g}_i}{\sum\limits_{in} \dot{n}_i \bar{h}_i - \sum\limits_{out} \dot{n}_i \bar{h}_i} \qquad (8.27)$$

This provides an estimate of the maximum efficiency attained *in the absence of irreversibilities*. The equation is often quoted in terms of the molar-based higher heating value (HHV) of the fuel thus:

$$\eta_{reaction} = \frac{\sum\limits_{in} \dot{n}_i \bar{g}_i - \sum\limits_{out} \dot{n}_i \bar{g}_i}{HHV} \qquad (8.28)$$

At room temperatures and pressures, $\Delta \bar{g} = -237.3 \, \text{kJ/mol}$ and $\text{HHV} = -285.8 \, \text{kJ/mol}$.

Therefore, for a pure hydrogen–oxygen fuel cell operating at standard temperature and pressure, this equation returns an impressive 83% efficiency. The reaction efficiency decreases with increasing temperature. At 1000 °C, the reaction efficiency drops to around 65%.

This is not the full story, however. Fuel cells rely on physical and chemical processes, for example, reaction and concentration kinetics, mixing, migration, absorption and dissolution, for their operation. These are neither instantaneous nor perfect.

From work considerations, the reversible electrical potential (E, volts) of the fuel cell is given by:

$$E = \frac{-\Delta G}{\dot{n}_e F} \tag{8.29}$$

where $\dot{n}_e$ is the number of kilomoles of valence electrons flowing in the external circuit and F is *Faraday's constant* (96.487 kJ/kmol V). For hydrogen, $n_e = 2$ per mol.

The *voltage efficiency* (η_{voltage}) of the fuel cell is expressed as the ratio of the actual voltage (V) appearing across the cell compared to the theoretical potential:

$$\eta_{\text{voltage}} = V/E \tag{8.30}$$

The voltage efficiency is inversely proportional to the load current.

Furthermore, an amount of fuel in excess of stoichiometric is usually supplied to the cell to compensate for poor fuel usage, and a third term, the *fuel utilization efficiency* (η_{fuel}), is used to describe its impact. It is expressed as the *reciprocal* of the ratio of actual fuel input ($n_{\text{fuel,act}}$) divided by the stoichiometric amount of fuel ($n_{\text{fuel,act}}/n_{\text{fuel,sto}}$). Then:

$$\eta_{\text{fuel}} = \frac{n_{\text{fuel,sto}}}{n_{\text{fuel,act}}} \tag{8.31}$$

8.10.3 Fuel Cell Types

The specifications of some of the more common fuel types are detailed in Table 8.4. The cell names usually reflect the electrolyte used. In all cases, the fuel is assumed to be hydrogen.

8.11 Energy Storage Technologies

Our electrical energy demand varies by month, by day, by hour and even by minute. This is a problem for power suppliers because, for reasons of stability, fossil fuel and nuclear power plants are not, in general, able easily to mirror a rapidly varying demand. Moreover, generating capacity must be available to cover the peak demand that may only occur for a small fraction of the day. Some of the criticisms aimed at renewable electrical energy forms are that they are unreliable, unpredictable (with the exception of tidal power) and exacerbate the problem of supply–demand matching. One solution being proposed is the construction of pan-continental super-grids, allowing surplus power capacity in one region of a continent to be routed to another experiencing its peak demand. Another potential solution is to generate electrical energy whenever available and store it until demanded.

Table 8.4 Types of fuel cell. Adapted from Hodge, B.K. (2010) *Alternative Energy Systems and Applications*, John Wiley & Sons, Inc.

Cell type	Ion carrier	Oxidant	Catalyst	Electrolyte	Operating temperature (°C)	Capacity (MW)	Efficiency (%)
Alkaline (AFC)	OH^-	O_2	Platinum/palladium	Aqueous potassium hydroxide	Up to 250	Up to 0.01	50–60
Phosphoric acid (PAFC)	H^+	Air	Finely dispersed platinum	Liquid phosphoric acid in silicon carbide matrix	~ 200	0.01–0.4	35–45
Molten carbonate (MCFC)	CO_2^{3-}	O_2	Nickel/nickel oxide	Molten lithium carbonate/potassium carbonate in porcus ceramic matrix	Up to 650	0.025–10	50–60
Solid oxide (SOFC)	O^{2-}	Air	Non-precious metals	Solid ceramic, e.g. zirconium oxide	Up to 1000	Up to 10	50–60
Proton or polymer exchange membrane (PEMFC)	H^+	Air	Platinum/ruthenium coating	Acid in thin plastic polymer membrane	Up to 90	Up to 0.025	40–50

8.11.1 Energy Storage Characteristics

The salient operating parameters when considering which form of energy store to employ are:

Energy density (J/m^3) – the energy stored per unit volume of store.
Specific energy content (J/kg) – the energy stored per unit mass of store.
Storage efficiency (%) – the fraction of the input energy that is usefully recovered.
Retrieval rate and depletion time (W, hrs) – how quickly and for how long the energy exchange can be maintained.
Capital and operating costs (£).

8.11.2 Energy Storage Technologies

There are a number of storage technologies which may suit a particular application. Final selection should be based on the criteria previously described.

8.11.2.1 Hydraulic Energy

Electrical energy can be stored by using it to increase the potential energy of water in a pumped storage scheme (see Figure 8.33). This requires a mountainous terrain, as excess electricity is used to pump water from a low-level lake or reservoir to another lake or reservoir at high elevation. At times of increased electricity demand, water is run back down to the lower reservoir via a turbine connected to an electrical generator. The pumping and generating functions are usually combined into one hydraulic machine (i.e. a pump/turbine).

The specific energy content is approximately 10 J/kg per metre elevation.

Potential energy, i.e. the energy associated with a mass (m) at a vertical distance (h, metres) above ground level in the earth's gravitational force field, is given by:

$$W_{\text{potential}} = mgh \tag{8.32}$$

Typical efficiencies are around 80%, with outputs in excess of 100 MW being brought on line and up to full power in seconds. Output can be maintained for hours, making the technology very useful for supply management purposes and smoothing out variations in renewable energy grid input.

The deployment of these systems is facilitated by using surplus energy from the grid at times of low power demand (usually at night) and cheap-tariff electricity.

The ratio of energy supplied to the network and the energy consumed during the pumping cycle must be considered to evaluate the overall viability of the scheme.

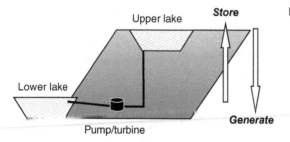

Figure 8.33 Pump storage scheme.

8.11.2.2 Pneumatic Energy

Energy can be stored by forcing the molecules of a gas into close proximity with each other, i.e. by compressing the gas and increasing its density in a compressed-air energy storage scheme (CAES).

When the pressure of the gas (typically 50–70 bar at 50 °C) is later reduced, the molecules will spring apart, releasing the stored energy.

In practical terms, excess electrical energy is used to compress air by forcing it into a closed pressure vessel. Large-scale pressure vessels are very expensive, and so natural cavities, like disused salt mines, have been used as containers. When electrical demand increases, the compressed air is released to a turbine generating electricity (see Figure 8.34).

If the compression is carried out isothermally, the energy of compression is given by:

$$W_{\text{comp}} = P_2 V_2 \left(\ln \frac{P_2}{P_1} \right) \tag{8.33}$$

where P_2 is the storage pressure and V_2 the storage volume.

Expanding the air directly during generation will result in a large temperature reduction. In order to protect the turbine, natural gas is used to reheat the air. Alternatively, the heat of compression may be used for this function if a thermal store is provided to operate in parallel with the pneumatic store.

Power outputs are of the order of 1–100's MW, with energy-depletion times in hours. Care must be taken with definitions of efficiency to include air cooling, drying, reheating of the air etc. Recent developments include cooling the air to its liquefaction point to reduce storage volume.

Again, the technology is very suitable for supply management.

8.11.2.3 Ionic Energy

On a small scale, electrical energy can be stored in rechargeable (or secondary) batteries.

Here, excess electrical energy is converted to electrochemical energy during charging and the reverse during discharge. Performance depends on the rate of charge/discharge.

Lead–acid batteries (specific energy content ~ 0.14 MJ/kg) have been used for many years for applications of < 10 MW and depletion times of around an hour.

Other materials are available, like nickel–iron and nickel–cadmium (specific energy content ~ 0.09 MJ/kg). Silica oxide-based batteries have energy storage capacities of up to 0.4 MJ/kg.

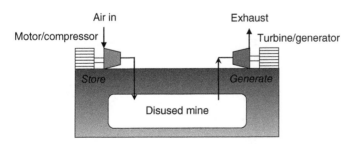

Figure 8.34 CAES scheme.

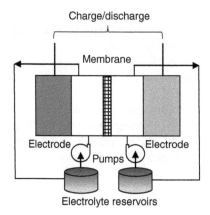

Figure 8.35 Flow battery schematic.

Newer systems like flow (pumped electrolyte) batteries (specific energy content ~ 1–3 MJ/kg) have better performance, and, with power outputs in the Megawatt range and energy depletion times of up to 10 hours, are being considered for supply management and renewable energy smoothing (see Figure 8.35).

8.11.2.4 Rotational Energy

Electrical energy can be stored by converting it to rotational kinetic energy in a flywheel. The amount of energy that can be stored depends on the mass of the flywheel, the square of its radius and the square of its rotational speed. So, a small change in speed will produce better returns than a small increase in mass. The kinetic energy associated with rotational motion (W_{rotate}, Joules) about an axis can be evaluated from:

$$W_{rotate} = \frac{1}{2}I_m\omega^2 \tag{8.34}$$

where I_m is the moment of inertia (kgm^2) that depends on the dimensions (radius, r; mass, m) and the position of the axis of rotation of the body in the relationship:

$$I_m = kmr^2 \tag{8.35}$$

The factor k for some simple geometries is given in Table 8.5.

The angular velocity of the body, ω (rad/s) can be determined with either knowledge of the time taken for one complete revolution of the body, t (seconds) or its rotational

Table 8.5 Geometric k factors.

Geometric shape	k
Rim-loaded thin ring	1
Uniform disc	1/2
Thin rectangular rod	1/2
Sphere	2/5
Spherical shell	2/3

speed in revolutions per minute (N), thus:

$$\omega = \frac{2\pi}{t} = \frac{2\pi N}{60} \tag{8.36}$$

Typical wheel peripheral speeds are 1500–2000 m/s or rotational speeds of 100 000 rpm. The maximum amount of energy that can be stored depends on the ratio of the strength of the flywheel material and its density. The best type of material would have high strength and low density, so, for example, carbon fibre (specific energy content ~ 0.77 MJ/kg) would make a better flywheel than steel (specific energy content ~ 0.17 MJ/kg). Storage efficiency (typically in the range 95–98%) is affected by friction losses, so flywheels often operate on magnetic bearings in a vacuum. Power outputs are limited to less than 1 MW and energy-depletion times are measured in minutes, putting them on the boundary for supply management.

8.11.2.5 Electrostatic Energy

A device capable of storing electrical energy in the form of an electric field is termed a *capacitor*. A capacitor comprises a material separating two plates capable of accumulating and maintaining an electric charge potential between them.

For a given voltage, V, the energy storage (W_{cap}, Joules) of a capacitor is given by:

$$W_{cap} = \frac{1}{2}CV^2 \tag{8.37}$$

where C (Farad) is the device capacitance.

The capacitance is a characteristic dependent on the material properties and dimensions of the device.

Super-capacitors and ultra-capacitors (specific energy content up to 0.03 MJ/kg) are electrical devices that charge much like a battery but discharge electrostatically. Storage efficiencies are high (98%) but power outputs of less than 100 kW are only possible for a few seconds, and so use is confined to maintaining the quality of a power supply.

8.11.2.6 Magnetic Energy

A device capable of storing electrical energy in the form of a magnetic field is termed an *inductor*. An inductor can be as simple as a current-carrying coil of wire.

For a given current, I (amps) the energy storage (W_{ind}, Joules) of an inductor is given by:

$$W_{ind} = \frac{1}{2}LI^2 \tag{8.38}$$

where L is the device self-inductance in Henrys.

For a simple inductor comprising a coil of wire wound around a structure or core, the inductance is dependent on the core dimensions and material as well as the number of turns of wire in the coil.

At very low temperatures (5–20 K), some metals become superconducting, i.e. they have no electrical resistance. Any current passed through a superconductor will produce a magnetic field capable of energy storage. Superconducting magnetic energy storage systems (SMESs) have energy densities of the order of 6.5 MJ/m^3. Power outputs in excess of capacitors are possible, but again their energy depletion timeframe is in seconds.

8.12 Worked Examples

Worked Example 8.1 – Nuclear fuel consumption
Estimate the mass of U-235 required per day to produce 1000 MW of electricity in a boiling-water reactor with an overall cycle efficiency of 33%.

Assume the fission of a U-235 atom yields 190 MeV of useful energy and that 85% of the neutrons are effective in the fission.

Take Avogadro's number to be 6.023×10^{23} neutrons/g mol, and 1 MeV $= 1.6 \times 10^{13}$ J.

Solution
Given: $\eta_{cycle} = 0.33$, $\dot{W}_e = 1000\,MW$.
Find m_{U235}.

Number of fissions required per Joule:

$$n_f = 1/(E_{U235} \times 1.6 \times 10^{13}) = 1/(190 \times 1.6 \times 10^{13}) = 3.3 \times 10^{10}/J$$

Thermal power input = electrical power output/overall cycle efficiency

$$= 1000/0.33$$
$$= 3030\,MW$$

Number of U-235 nuclei used:

$$N_{U235} = \frac{n_f \times \frac{seconds}{day} \times thermal\,power\,input}{0.85}$$

$$= \frac{3.3 \times 10^{10} \times 86\,400 \times 3030 \times 10^6}{0.85} = 10.1 \times 10^{24}$$

Therefore, the mass of U-235 used per day:

$$m_{U235} = \frac{N_{U235} \times M_{U235}}{Avogadro's\,number}$$

$$m_{U235} = \frac{10.1 \times 10^{24} \times 235}{6.023 \times 10^{23}} = 3.953\ kg/day.$$

Worked Example 8.2 – PV module power output
A PV array comprises two parallel circuits, each containing two PV modules in series. Calculate the maximum power available from each module and from the full PV array if the single module's current is 3.75 amp and 12 V.

Comment on the result found and discuss possible reasons for not achieving that value.

Solution
For a *single module*, the maximum power at full sun can be calculated as:

$$\dot{W}_{max,collector} = V_{single} \times I_{single} = 12 \times 3.75 = 45\,W$$

The output voltage from the array is the sum of the series connection of the PV panels:
$V_{tot} = 24$ volts.

The output current will be equal to the sum of the parallel branch currents. Since each PV panel produces 3.75 amperes at full sun, the total current (I_{tot}) will be 7.5 amperes. Then, the maximum power of the photovoltaic array at full sun can be calculated as:

$$\dot{W}_{max,array} = V_{tot} \times I_{tot} = 24 \times 7.5 = 180 \, \text{W}.$$

The PV array reaches its maximum of 180 watts in full sun because the maximum power output of each PV panel or module is equal to 45 watts (12 volts × 3.75 amps). However, due to different levels of solar radiation, temperature effect, electrical losses etc., the real maximum output power is usually a lot less than the calculated 180 watts.

Worked Example 8.3 – Solar thermal power production

A solar thermal collector whose active surface is 5 m by 2 m is exposed to a light intensity of 1000 watts per square metre.

(a) Calculate the potential thermal energy available if the unit is operating at 20% efficiency.
(b) If water at 20 °C is fed into the collector and is heated to 50 °C, calculate the amount of water it can heat. Assume $C_p = 4.2$ kJ/kg K.

Solution

$$A_{collector} = 5 \times 2 = 10 \, \text{m}^2$$

Incident solar irradiance (W) = Light intensity (W/m^2) x Collector area (m^2)

$$= 1000 \times 10 = 10\,000 \, \text{W}$$

Electrical power generation (W) = Incident solar irradiance (W) × Efficiency (%)

$$= 10\,000 \times 0.2 = 2000 \, \text{W} = 2 \, \text{kW}$$

For the water side:

$$\dot{Q} = \dot{m} \times C_p \times \Delta T$$

$$\dot{m} = \frac{\dot{Q}}{C_p \times (T_2 - T_1)} = \frac{2}{4.2 \times (50 - 10)} = 0.0158 \, \text{kg/s} \quad \text{or} \quad 57 \, \text{kg/hr}.$$

Worked Example 8.4 – Wind turbine selection

The manufacturers' specifications for two HAWTs (A and B) are shown below:

Item	Turbine A	Turbine B
Rotor diameter (m)	25	28
Power coefficient	0.38	0.35
Gearbox efficiency (%)	90	88
Generator efficiency (%)	98	95
Capital cost (£)	99 000	103 000
Maintenance cost/year (£)	4000	4000

(a) Draw up a table for the performance of each turbine for wind speeds of 4–2 m/s in intervals of 2 m/s.
(b) Assuming the site wind availability to be 2000 hours per year at an average wind speed of 6 m/s, select the wind turbine that will provide the best investment.

Assume life expectancy for each turbine to be 20 years and that the electricity generated can be sold at 6 pence per kWh, remaining constant over the term.
Take the air density to be 1.2 kg/m³.

Solution
(a) Calculate the shaft power of each wind rotor using:

$$\dot{W}_{wind} = \frac{1}{2}C_p\rho A \bar{V}^3 \eta_{gb}\eta_{gen}$$

Wind speed m/s	Turbine A Power (kW) = 98.713 V^3	Turbine B Power (kW) = 108.101 V^3
4	6.317	6.918
6	21.322	23.350
8	50.541	55.348
10	98.713	108.101
12	170.576	186.799

(b) At 6 m/s, Turbine B, produces more power.
Annual difference in energy generation (B − A) = (23.350 − 21.322) × 2000

= 4056 kWh/annum

Value of energy generated (B − A) over 20 years = 4056 × 0.06 × 20 = £4867
Hence, Turbine B should be selected, even though its capital cost is £4000 in excess of A, as it will result in an extra income of £867 over the term.

Worked Example 8.5 – Hydropower generation
In a hydropower plant, a water reservoir is located 55 m above a turbine house. The penstock diameter is 1.2 m. Assuming frictional losses to be negligible, determine:

(a) The free velocity (m/s) of water emerging at the turbine inlet.
(b) The maximum electricity generation (MW) if the hydraulic turbine efficiency is 80% and the electrical generator's efficiency is 90%.

Solution
Given: $h = 55$ m, $\Delta h_f = 0$, $D_{pen} = 1.2$ m, $\eta_t = 0.8$, $\eta_{gen} = 0.9$.
Find $\bar{V}_{pen}$, $\dot{W}_{elec}$.

(a) The potential energy behind the dam is converted into kinetic energy once the water is allowed through the pipe falling to the generator house, hence:

$$m \times g \times h = \frac{1}{2} \times m \times \bar{V}^2$$

$$\bar{V} = \sqrt{2.g.h}$$

$$= \sqrt{2 \times 9.81 \times 55}$$

$$= 32.8\,\text{m/s}$$

Using the continuity equation:

$$\dot{V} = A \times \bar{V} = \frac{\pi}{4} \times 1.2^2 \times 32.8 = 37.152\,\text{m}^3/\text{s}.$$

(b) Electrical power

The maximum electrical power generated can be determined by:

$$\dot{W}_{\text{elec}} = \rho g(h - \Delta h_{\text{f}})\dot{V}\eta_t\eta_{\text{gen}}$$

$$= 1000 \times 9.81 \times (55 - 0) \times 37.152 \times 0.8 \times 0.9$$

$$= 14.4 \times 10^6\,\text{W} = 14.4\,\text{MW}.$$

Worked Example 8.6 – Determination of wave power production

Determine the possible energy available from ocean swells with a wave height of 2 m and a wave energy period of 10 seconds.

Assume that the density of sea water is 1020 kg/m^3.

Solution

Given: $H = 2\,\text{m}$, $t = 10$ seconds, $\rho = 1020$ kg/m^3.

Find $\dot{W}_{\text{wave}}$.

Wave power is determined from:

$$\dot{W}_{\text{wave}} = \frac{\rho g^2}{32\pi} H^2 T$$

$$= \frac{1020 \times 9.81^2}{32\pi} \times 2^2 \times 10$$

$$= 39\,000\,\text{W/m}^2$$

$$= 39\,\text{kW/m}^2.$$

Worked Example 8.7 – Thermoelectric energy from waste

A thermoelectric generator (TEG) is attached to a waste heat pipe with a contact temperature of 210 °C. The TEG has the specification shown below. If the heat sink temperature is 10 °C, determine the potential power generation capacity (W) of the device.

Specification

Length: 20 mm
Cross-sectional area: 0.004 m^2
Seebeck coefficient: 0.1 V/K
Resistivity: 1 Ω/m

Solution

Given: $L = 0.02$ m, $A = 0.004$ m^2, $\alpha = 0.1$ V/K, $\rho = 1\Omega$/m, $T_{hot} = 210\,°$C, $T_{cold} = 10\,°$C.

Find $\dot{W}_{TE}$.

$$\Delta T = 210 - 10 = 200\,°C$$

The total power generated by a thermoelectric device is given by:

$$\dot{W}_{TE} = \Delta T^2 \frac{\alpha^2 A}{\rho L}$$

$$= (210 - 10)^2 \times \frac{0.1^2 \times 0.004}{1 \times 0.02} = 80\,W.$$

Worked Example 8.8 – Flywheel energy storage

A flywheel weighing 100 kg takes the form of a uniform solid disc with a radius of 0.3 m and a geometric factor $k = 0.5$.

Determine the energy storage capacity (MJ) of the wheel if rotated at 20 000 rpm.

Solution

Given: $m = 100$ kg , $r = 0.3$ m, $N = 20\,000$ rpm, $k = 0.5$.

Find W_{rotate}.

$$\omega = 2\pi N/60 = 2 \times 3.142 \times \frac{20\,000}{60} = 2095\ \text{rad/s}$$

$$I_m = kmr^2 = 0.5 \times 100 \times 0.3^2 = 4.5\ \text{kg/m}^2$$

$$W_{rotate} = \frac{1}{2}I_m\omega^2 = 0.5 \times 4.5 \times 2095^2 = 9.88 \times 10^6 J = 9.88\ \text{MJ}.$$

Worked Example 8.9 – Sizing a compressed-air energy store

A pneumatic energy store, also known as 'compressed-air storage' is required to supply 10 MW for 1 hr.

If the energy density of the store is 18 MJ/m^3, determine the necessary store volume (m^3). Take 1 kWh = 3.6 MJ.

Solution

Energy storage requirement $= 10 \times 1$

$$= 10\ \text{MWh}$$

$$= 10 \times 3600 = 36\,000\ \text{MJ}.$$

Store volume = Energy storage requirement/store energy density

$$= 36\,000/18$$

$$= 2000\ \text{m}^3$$

Worked Example 8.10 – Capacitive and inductive energy storage

Compare the energy storage capacity (MJ) of the two electrical systems detailed below:

(a) An inductor with a self-inductance of 10 μHenrys and a supply current of 13 A.

(b) A super-capacitor with a capacitance of 1 μF and a voltage of 2.7 V.

Solution

For the inductor:

Given: $L = 10 \times 10^{-6}$ H, $I = 13$ A.

Find W_{ind}.

$$W_{ind} = \frac{1}{2}LI^2 = \frac{1}{2} \times 10 \times 10^{-6} \times 13^2 = 8.45 \times 10^{-4} \text{ J.}$$

For the capacitor:

Given: $C = 1 \times 10^{-6}$ F, $V = 2.7$ V.

Find W_{cap}.

$$W_{cap} = \frac{1}{2}CV^2 = \frac{1}{2} \times 1 \times 10^{-6} \times 2.7^2 = 3.645 \times 10^{-6} \text{ J.}$$

8.13 Tutorial Problems

8.1 Estimate the thermal energy liberated by 1 kg/s of U-235 in a nuclear reactor. Assume the fission of a U-235 atom yields 190 MeV of useful energy and that 85% of the neutrons are effective in the fission. Take Avogadro's number as 6.023×10^{23} neutrons/g mol, and 1 MeV $= 1.6 \times 10^{13}$ J.

[Answer: 766 kW]

8.2 On a bright day at a latitude of 52° North, a solar PV cell of dimension 1 m × 1 m will generate a current of 2 A and a voltage of 5 V.

(a) What power (W) is being generated per cell?

(b) How many such PV cells are required to generate 1 kW?

[Answers: 10 W, 100 cells]

8.3 A solar thermal collector whose active surface is 2 m by 2 m is exposed to a light intensity of 1000 watts/m².

(a) Calculate the potential thermal power (kW) available if the unit is operating at 20% efficiency.

(b) Water is supplied to the collector at 20 °C. Determine the maximum flow rate (kg/hr) through the collector if the requirement is to provide water at 50 °C. Assume $C_p = 4.2$ kJ/kg K.

[Answers: 0.8 kW, 22.86 kg/hour]

8.4 A two-bladed horizontal-axis wind generator has a rotor diameter of 20 metres. Operating at a wind speed of 8.1 metres per second, the rotor extracts 35% of the energy of the wind. If the efficiency of the generator in the power train is 85%, find the electrical power output (kW). Take the density of air to be 1.23 kg/m³.

[Answer: 30.5 kW]

8.5   A small-scale hydropower scheme has a gross head of 125 m at a flow rate of 90 litres per second. If the transmission pipework is designed to limit the head loss to 10% of gross head, using the data below, estimate the electrical power output (MW) for the scheme.

Data
Electrical generator efficiency 85%; turbine efficiency 80%.

[Answer: 67.54 kW]

8.6   Determine the possible energy available from ocean swells with a wave height of 4 m and a wave energy period of 10 seconds.
Assume that the density of water is 1020 kg/m^3.

[Answer: 156 kW/m^2]

8.7   A waste heat pipe with a contact temperature of 420 °C is used as the heat source for a thermoelectric generator. The TEG has the specification shown below. If the heat sink temperature is 10 °C, determine the potential power generation capacity (W) of the device.

Specification
Length: 20 mm
Cross-sectional area: 0.004 m^2
Seebeck coefficient: 0.1 V/K
Resistivity: 1 Ω/m.

[Answer: 336 W]

8.8   Determine the energy storage capacity (MJ) of a flywheel comprising a uniform circular disc of 0.5 m radius and 100 kg weight when rotating at 10 000 rpm. Assume $k = 0.5$.

[Answer: 6.851 MJ]

8.9   A pneumatic energy air store has ambient air compressed isothermally to 60 bar. If the store volume is 200 000 m^3, determine the energy stored (MWh).

[Answer: 136.4 MWh]

8.10   Which one of the following devices has a greater energy storage capacity?
(a) An inductor with a self-inductance of 20 μHenrys and a current of 13 A.
(b) A super-capacitor with a capacitance of 2 μF and an applied voltage of 2.7 V.

[Answer: 1.69 mJ, 7.29 μJ, the inductor (a) produces more]

A

Properties of Water and Steam

The tables in this appendix are adapted from Moran, Shapiro, Boettner and Bailey (2012) *Principles of Engineering Thermodynamics*, 7th edition, John Wiley & Sons.

T	P = 0.01 MPa (45.81 °C)		
	v	h	s
°C	m³/kg	kJ/kg	kJ/kgK
Sat liq	0.00101	191.8	0.649
Sat vap	14.674	2584.7	8.1502
50	14.869	2592.6	8.1749
100	17.196	2687.5	8.4479
150	19.512	2783	8.6882
200	21.825	2879.5	8.9038
250	24.136	2977.3	9.1002
300	26.445	3076.5	9.2813
400	31.063	3279.6	9.6077
500	35.679	3489.1	9.8978
600	40.295	3705.4	10.1608
700	44.911	3928.7	10.4028
800	49.526	4159	10.6281

T	P = 0.05 MPa (81.33 °C)		
	v	h	s
°C	m³/kg	kJ/kg	kJ/kgK
Sat liq	0.00103	340.5	1.091
Sat vap	3.240	2645.9	7.5939
50	-	-	-
100	3.418	2682.5	7.6947
150	3.889	2780.1	7.9401
200	4.356	2877.7	8.158
250	4.820	2976	8.3556
300	5.284	3075.5	8.5373
400	6.209	3278.9	8.8642
500	7.134	3488.7	9.1546
600	8.057	3705.1	9.4178
700	8.981	3928.5	9.6599
800	9.904	4158.9	9.8852

T	P = 0.10 MPa (99.63 °C)		
	v	h	s
°C	m³/kg	kJ/kg	kJ/kgK
Sat liq	0.00104	417.4	1.303
Sat vap	1.694	2675.5	7.3594
100	1.6958	2676.2	7.3614
150	1.9364	2776.4	7.6143
200	2.172	2875.3	7.8343
250	2.406	2974.3	8.0333
300	2.639	3074.3	8.2158
400	3.103	3278.2	8.5435
500	3.565	3488.1	8.8342
600	4.028	3704.4	9.0976
700	4.49	3928.2	9.3398
800	4.952	4158.6	9.5652

T	P = 0.2 MPa (120.23 °C)		
	v	h	s
°C	m³/kg	kJ/kg	kJ/kgK
Sat liq	0.00106	504.7	1.53
Sat vap	0.8857	2706.7	7.1272
100	-	-	-
150	0.9596	2768.8	7.2795
200	1.0803	2870.5	7.5066
250	1.1988	2971	7.7086
300	1.3162	3071.8	7.8926
400	1.5493	3276.6	8.2218
500	1.7814	3487.1	8.5133
600	2.013	3704	8.777
700	2.244	3927.6	9.0194
800	2.475	4158.2	9.2449

Conventional and Alternative Power Generation: Thermodynamics, Mitigation and Sustainability, First Edition. Neil Packer and Tarik Al-Shemmeri.
© 2018 John Wiley & Sons Ltd. Published 2018 by John Wiley & Sons Ltd.

T	P = 0.30 MPa (133.55 °C)		
	v	h	s
°C	m³/kg	kJ/kg	kJ/kgK
Sat liq	0.00107	561.5	1.672
Sat vap	0.6058	2725.3	6.9919
150	0.6339	2761	7.0778
200	0.7163	2865.6	7.3115
250	0.7964	2967.6	7.5166
300	0.8753	3069.3	7.7022
400	1.0315	3275	8.033
500	1.1867	3486	8.3251
600	1.3414	3703.2	8.5892
700	1.4957	3927.1	8.8319
800	1.6499	4157.8	9.0576

T	P = 0.40 MPa (143.63 °C)		
	v	h	s
°C	m³/kg	kJ/kg	kJ/kgK
Sat liq	0.00108	604.7	1.7766
Sat vap	0.4625	2738.6	6.8959
150	0.4708	2752.8	6.9299
200	0.5342	2860.5	7.1706
250	0.5951	2964.2	7.3789
300	0.6548	3066.8	7.5662
400	0.7726	3273.4	7.8985
500	0.8893	3484.9	8.1913
600	1.0055	3702.4	8.4558
700	1.1215	3926.5	8.6987
800	1.2372	4157.3	8.9244

T	P = 0.50 MPa (151.86 °C)		
	v	h	s
°C	m³/kg	kJ/kg	kJ/kgK
Sat liq	0.00109	640.2	1.8607
Sat vap	0.3749	2748.7	6.8213
200	0.4249	2855.4	7.0592
250	0.4744	2960.7	7.2709
300	0.5226	3064.2	7.4599
350	0.5701	3167.7	7.6329
400	0.6173	3271.9	7.7938
500	0.7109	3483.9	8.0873
600	0.8041	3701.7	8.3522
700	0.8969	3925.9	8.5952
800	0.9896	4156.9	8.8211

T	P = 0.60 MPa (158.85 °C)		
	v	h	s
°C	m³/kg	kJ/kg	kJ/kgK
Sat liq	0.0011	670.6	1.9312
Sat vap	0.3175	2756.8	6.7600
200	0.3520	2850.1	6.9665
250	0.3938	2957.2	7.1816
300	0.4344	3061.6	7.3724
350	0.4742	3165.7	7.5464
400	0.5137	3270.3	7.7079
500	0.592	3482.8	8.0021
600	0.6697	3700.9	8.2674
700	0.7472	3925.3	8.5107
800	0.8245	4156.6	8.7367

T	P = 0.80 MPa (170.43 °C)		
	v	h	s
°C	m³/kg	kJ/kg	kJ/kgK
Sat liq	0.00111	721.1	2.0462
Sat vap	0.2404	2769.1	6.6628
200	0.2608	2839.3	6.8158
250	0.2931	2950	7.0384
300	0.3241	3056.5	7.2328
350	0.3544	3161.7	7.4089
400	0.3843	3267.1	7.5716
500	0.4433	3480.6	7.8673
600	0.5018	3699.4	8.1333
700	0.5601	3924.2	8.377
800	0.6181	4155.6	8.6033

T	P = 1.0 MPa (179.91 °C)		
	v	h	s
°C	m³/kg	kJ/kg	kJ/kgK
Sat liq	0.00113	762.8	2.1387
Sat vap	0.1944	2778.1	6.5865
200	0.206	2827.9	6.694
250	0.2327	2942.6	6.9247
300	0.2579	3051.2	7.1229
350	0.2825	3157.7	7.3011
400	0.3066	3263.9	7.4651
500	0.3541	3478.5	7.7622
600	0.4011	3697.9	8.029
700	0.4478	3923.1	8.2731
800	0.4943	4154.7	8.4996

T	P = 1.20 MPa (187.99 °C)		
	v	**h**	**s**
°C	m³/kg	kJ/kg	kJ/kgK
Sat liq	0.00114	798.6	2.2166
Sat vap	0.1633	2784.4	6.5233
200	0.1693	2815.9	6.5898
250	0.1923	2935	6.8294
300	0.2138	3045.8	7.0317
350	0.2345	3153.6	7.2121
400	0.2548	3260.7	7.3774
500	0.2946	3476.3	7.6759
600	0.3339	3696.3	7.9435
700	0.3729	3922	8.1881
800	0.4118	4153.8	8.4148

T	P = 1.4 MPa (195.07 °C)		
	v	**h**	**s**
°C	m³/kg	kJ/kg	kJ/kgK
Sat liq	0.00115	830.3	2.2842
Sat vap	0.1408	2790	6.4693
200	0.1430	2803.3	6.4975
250	0.1635	2927.2	6.7467
300	0.1823	3040.4	6.9534
350	0.2003	3149.5	7.1360
400	0.2178	3257.5	7.3026
500	0.2521	3474.1	7.6027
600	0.2860	3694.8	7.8710
700	0.3195	3920.8	8.1160
800	0.3528	4153	8.3431

T	P = 1.6 MPa (201.41 °C)		
	v	**h**	**s**
°C	m³/kg	kJ/kg	kJ/kgK
Sat liq	0.00116	858.8	2.3442
Sat vap	0.1238	2794	6.4218
225	0.1329	2857.3	6.5518
250	0.1418	2919.2	6.6732
300	0.1586	3034.8	6.8844
350	0.1746	3145.4	7.0694
400	0.1900	3254.2	7.2374
500	0.2203	3472	7.539
600	0.2500	3693.2	7.808
700	0.2794	3919.7	8.0535
800	0.3086	4152.1	8.2808

T	P = 1.8 MPa (207.15 °C)		
	v	**h**	**s**
°C	m³/kg	kJ/kg	kJ/kgK
Sat liq	0.00116	885	2.398
Sat vap	0.11042	2797.1	6.3794
225	0.11673	2846.7	6.4808
250	0.12497	2911	6.6066
300	0.14021	3029.2	6.8226
350	0.15457	3141.2	7.01
400	0.16847	3250.9	7.1794
500	0.1955	3469.8	7.4825
600	0.2220	3691.7	7.7523
700	0.2482	3918.5	7.9983
800	0.2742	4151.2	8.2258

T	P = 2.0 MPa (212.42 °C)		
	v	**h**	**s**
°C	m³/kg	kJ/kg	kJ/kgK
Sat liq	0.00117	908.8	2.4474
Sat vap	0.09963	2799.5	6.3409
225	0.1038	2835.8	6.4147
250	0.1114	2902.5	6.5453
300	0.1255	3023.5	6.7664
350	0.1386	3137	6.9563
400	0.1512	3247.6	7.1271
500	0.1757	3467.6	7.4317
600	0.1996	3690.1	7.7024
700	0.2232	3917.4	7.9487
800	0.2467	4150.3	8.1765

T	P = 3.0 MPa (233.90 °C)		
	v	**h**	**s**
°C	m³/kg	kJ/kg	kJ/kgK
Sat liq	0.001216	1008.4	2.6457
Sat vap	0.06668	2804.2	6.1869
250	0.07058	2855.8	6.2872
300	0.08114	2993.5	6.539
350	0.09053	3115.3	6.7428
400	0.09936	3230.9	6.9212
450	0.10787	3344	7.0834
500	0.11619	3456.5	7.2338
600	0.13243	3682.3	7.5085
700	0.14838	3911.7	7.7571
800	0.16414	4145.9	7.9862

T	P = 3.5 MPa (242.60 °C)		
	v	**h**	**s**
°C	m³/kg	kJ/kg	kJ/kgK
Sat liq	0.00123	1050	2.757
Sat vap	0.0507	2803.4	6.1253
250	0.05872	2829.2	6.1749
300	0.06842	2977.5	6.4461
350	0.07678	3104	6.6579
400	0.08453	3222.3	6.8405
450	0.09196	3337.2	7.0052
500	0.09918	3450.9	7.1572
600	0.11324	3678.4	7.4339
700	0.12699	3908.8	7.6837
800	0.14056	4143.7	7.9134

T	P = 4.0 MPa (250.4 °C)		
	v	**h**	**s**
°C	m³/kg	kJ/kg	kJ/kgK
Sat liq	0.00125	1087.3	2.7964
Sat vap	0.04978	2801.4	6.0701
275	0.05457	2886.2	6.2285
300	0.05884	2960.7	6.3615
350	0.06645	3092.5	6.5821
400	0.07341	3213.6	6.769
450	0.08002	3330.3	6.9363
500	0.08643	3445.3	7.0901
600	0.09885	3674.4	7.3688
700	0.11095	3905.9	7.6198
800	0.12287	4141.5	7.8502

T	P = 4.5 MPa (257.49 °C)		
	v	**h**	**s**
°C	m³/kg	kJ/kg	kJ/kgK
Sat liq	0.00127	1120	2.86
Sat vap	0.04406	2798.3	6.0198
275	0.0473	2863.2	6.1401
300	0.05135	2943.1	6.2828
350	0.0584	3080.6	6.5131
400	0.06475	3204.7	6.7047
450	0.07074	3323.3	6.8746
500	0.07651	3439.6	7.0301
600	0.08765	3670.5	7.311
700	0.09847	3903	7.5631
800	0.10911	4139.3	7.7942

T	P = 5.0 MPa (263.99 °C)		
	v	**h**	**s**
°C	m³/kg	kJ/kg	kJ/kgK
Sat liq	0.001286	1154.2	2.921
Sat vap	0.03944	2794.3	5.9734
275	0.04141	2838.3	6.0544
300	0.04532	2924.5	6.2084
350	0.05194	3068.4	6.4493
400	0.05781	3195.7	6.6459
450	0.0633	3316.2	6.8186
500	0.06857	3433.8	6.9759
600	0.07869	3666.5	7.2589
700	0.08849	3900.1	7.5122
800	0.09811	4137.1	7.744

T	P = 6.0 MPa (257.64 °C)		
	v	**h**	**s**
°C	m³/kg	kJ/kg	kJ/kgK
Sat liq	0.00132	1213.3	3.0267
Sat vap	0.03244	2784.3	5.8892
300	0.03616	2884.2	6.0674
350	0.04223	3043	6.3335
400	0.04739	3177.2	6.5408
450	0.05214	3301.8	6.7193
500	0.05665	3422.2	6.8803
550	0.06101	3540.6	7.0288
600	0.06525	3658.4	7.1677
700	0.07352	3894.2	7.4234
800	0.0816	4132.7	7.6566

T	P = 7.0 MPa (286.88 °C)		
	v	**h**	**s**
°C	m³/kg	kJ/kg	kJ/kgK
Sat liq	0.00135	1267	3.1211
Sat vap	0.02737	2772.1	5.8133
300	0.02947	2838.4	5.9305
350	0.03524	3016	6.2283
400	0.03993	3158.1	6.4478
450	0.04416	3287.1	6.6327
500	0.04814	3410.3	6.7975
550	0.05195	3530.9	6.9486
600	0.05565	3650.3	7.0894
700	0.06283	3888.3	7.3476
800	0.06981	4128.2	7.5822

T	P = 8.0 MPa (295.06 °C)		
	v	h	s
°C	m³/kg	kJ/kg	kJ/kgK
Sat liq	0.00138	1316.7	3.2068
Sat vap	0.02352	2758	5.7432
300	0.02426	2785	5.7906
350	0.02995	2987.3	6.1301
400	0.03432	3138.3	6.3634
450	0.03817	3272	6.5551
500	0.04175	3398.3	6.724
550	0.04516	3521	6.8778
600	0.04845	3642	7.0206
700	0.05481	3882.4	7.2812
800	0.06097	4123.8	7.5173

T	P = 9.0 MPa (303.4 °C)		
	v	h	s
°C	m³/kg	kJ/kg	kJ/kgK
Sat liq	0.001418	1363.3	3.2858
Sat vap	0.02048	2742.1	5.6772
350	0.0258	2956.6	6.0361
400	0.02993	3117.8	6.2854
450	0.0335	3256.6	6.4844
500	0.03677	3386.1	6.6576
550	0.03987	3511	6.8142
600	0.04285	3633.7	6.9589
650	0.04574	3755.3	7.0943
700	0.04857	3876.5	7.2221
800	0.05409	4119.3	7.4596

T	P = 10.0 MPa (311.06 °C)		
	v	h	s
°C	m³/kg	kJ/kg	kJ/kgK
Sat liq	0.001452	1407.6	3.356
Sat vap	0.018026	2724.7	5.6141
325	0.019861	2809.1	5.7568
350	0.02242	2923.4	5.9443
400	0.02641	3096.5	6.212
450	0.02975	3240.9	6.419
500	0.03279	3373.7	6.5966
550	0.03564	3500.9	6.7561
600	0.03837	3625.3	6.9029
650	0.04101	3748.2	7.0398
700	0.04358	3870.5	7.1687
800	0.04859	4114.8	7.4077

T	P = 12.0 MPa (324.68 °C)		
	v	h	s
°C	m³/kg	kJ/kg	kJ/kgK
Sat liq	0.001526	1494.86	3.501
Sat vap	0.014269	2682.34	5.485
325	0.0143	2687.90	5.495
350	0.0172	2842.10	5.748
400	0.0211	3048.35	6.067
450	0.0242	3206.90	6.294
500	0.0268	3347.43	6.482
550	0.0293	3479.35	6.647
600	0.0317	3606.79	6.798
650	0.0339	3731.89	6.937
700	0.0361	3855.90	7.068
800	0.0404	4103.48	7.310

T	P = 15.0 MPa (342.24 °C)		
	v	h	s
°C	m³/kg	kJ/kg	kJ/kgK
Sat liq	0.00165	1610	3.685
Sat vap	0.01034	2610.5	5.3098
350	0.01147	2692.4	5.4421
400	0.01565	2975.5	5.8811
450	0.01845	3156.2	6.1404
500	0.02080	3308.6	6.3443
550	0.02293	3448.6	6.5199
600	0.02491	3582.3	6.6776
650	0.02680	3712.3	6.8224
700	0.02861	3840.1	6.9572
800	0.03210	4092.4	7.204

T	P = 16.0 MPa (347.36 °C)		
	v	h	s
°C	m³/kg	kJ/kg	kJ/kgK
Sat liq	0.00171	1653.58	3.751
Sat vap	0.00931	2577.11	5.238
350	0.0098	2606.12	5.285
400	0.0143	2943.90	5.809
450	0.0170	3136.63	6.086
500	0.0193	3294.36	6.297
550	0.0214	3436.94	6.476
600	0.0232	3571.77	6.635
650	0.0250	3702.34	6.780
700	0.0267	3830.58	6.916
800	0.0301	4084.30	7.164

T	P = 20.0 MPa (365.81 °C)		
	v	h	s
°C	m³/kg	kJ/kg	kJ/kgK
Sat liq	0.00204	1826.3	4.0139
Sat vap	0.00583	2409.7	4.9269
400	0.00994	2818.1	5.554
450	0.0127	3060.1	5.9017
500	0.01477	3238.2	6.1401
550	0.01656	3393.5	6.3348
600	0.01818	3537.6	6.5048
650	0.01969	3675.3	6.6582
700	0.02113	3809	6.7993
800	0.02385	4069.7	7.0544

T	P = 25.0 MPa		
	v	h	s
°C	m³/kg	kJ/kg	kJ/kgK
375	0.001973	1848	4.032
400	0.006004	2580.2	5.1418
425	0.007881	2806.3	5.4723
450	0.009162	2949.7	5.6744
500	0.011123	3162.4	5.9592
550	0.012724	3335.6	6.1765
600	0.014137	3491.4	6.3602
650	0.015433	3637.4	6.5229
700	0.016646	3777.5	6.6707
800	0.018912	4047.1	6.9345

B

Thermodynamic Properties of Fuels and Combustion Products

Table B.1 Enthalpy of formation, reaction and HHVs for selected compounds at 25 °C, 1 atm.

Substance	Formula	Molar mass M (kg/kmol)	$\bar{h}_f^0$ kJ/kmol	Higher heating value HHV (kJ/kg)	$\bar{h}_r$ kJ/kmol
Carbon	C (s)	12.01	0	32 770	−393 600
Hydrogen	H_2 (g)	2.016	0	141.789	−285 800
Nitrogen	N_2 (g)	28.01	0	—	—
Oxygen	O_2 (g)	32	0	—	—
Carbon monoxide	CO (g)	28.01	−110 530	—	—
Carbon dioxide	CO_2 (g)	44.01	−393 520	—	—
Water	H_2O (g)	18.02	−241 820	—	—
Water	H_2O (l)	18.02	−285 830	—	—
Hydrogen peroxide	H_2O_2 (g)	34.02	−136 310	—	—
Ammonia	NH_3 (g)	17.03	−46 190	—	—
Methane	CH_4 (g)	16.04	−74 850	55 510	−890 400
Acetylene	C_2H_2 (g)	26.04	+226 730	49 910	−1 299 700
Ethylene	C_2H_4 (g)	28.05	+52 280	50 300	−1 411 000
Ethane	C_2H_6 (g)	30.07	−84 680	51 870	−1 599 900
Propylene	C_3H_6 (g)	42.08	+20 410	48 920	−2 058 600
Propane	C_3H_8 (g)	44.09	−103 850	50 350	−2 220 000
Butane	C_4H_{10} (g)	58.12	−126 150	49 500	−2 877 000
Pentane	C_5H_{12} (g)	72.15	−146 440	49 010	−3 536 100
Octane	C_8H_{18} (g)	114.22	−208 450	48 260	−5 512 300
Octane	C_8H_{18} (l)	114.22	−249 910	47 900	−5 471 100
Benzene	C_6H_6 (g)	78.11	+82 930	42 270	−3 301 700
Methyl alcohol	CH_3OH (g)	32.04	−200 670	23 850	−764 200

(Continued)

Conventional and Alternative Power Generation: Thermodynamics, Mitigation and Sustainability,
First Edition. Neil Packer and Tarik Al-Shemmeri.
© 2018 John Wiley & Sons Ltd. Published 2018 by John Wiley & Sons Ltd.

Table B.1 (Continued)

Substance	Formula	Molar mass M (kg/kmol)	$\bar{h}_f^0$ kJ/kmol	Higher heating value HHV (kJ/kg)	$\bar{h}_r$ kJ/kmol
Methyl alcohol	CH_3OH (l)	32.04	−238 810	22 670	−726 300
Ethyl alcohol	C_2H_5OH (g)	46.07	−235 310	30 590	−1 409 300
Ethyl alcohol	C_2H_5OH (l)	46.07	−277 690	29 670	−1 366 900

The enthalpy of reaction $\left(\bar{h}_r\right)$ is approximated from HHVs.

Source: Adapted from Moran, Shapiro, Boettner and Bailey (2012) *Principles of Engineering Thermodynamics*, 7th edition, John Wiley & Sons.

Table B.2 Some temperature-dependent gaseous enthalpies.

T (K)	CO_2	H_2O	N_2	O_2	CO
0	−9364	−9904	−8669	−8682	−8699
100	−6456	−6615	−5770	−5778	−5770
200	−3414	−3280	−2858	−2866	−2858
298.15	0	0	0	0	0
300	67	63	54	54	54
400	4008	3452	2971	3029	2975
600	12 916	10 498	8 891	9 247	10 196
800	22 815	17 991	15 046	15 841	15 175
1000	33 405	25 978	21 460	22 707	21 686
1200	44 484	34 476	28 108	29 765	28 426
1400	55 907	43 447	34 396	36 966	35 338
1600	67 580	52 844	41 903	44 279	42 384
1800	79 442	62 609	48 982	51 689	49 522
2000	91 450	72 689	56 141	59 199	56 739
2200	103 570	83 036	63 371	66 802	64 019
2400	115 790	93 604	70 651	74 492	71 346
2600	128 080	104 370	77 981	82 274	78 714
2800	140 440	115 290	85 345	90 144	86 115
3000	152 860	126 360	92 738	98 098	93 542

$\bar{h} - \bar{h}^0$ (kJ/kmol) at 1 atm.

Source: Adapted from Moran, Shapiro, Boettner and Bailey (2012) *Principles of Engineering Thermodynamics*, 7th edition, John Wiley & Sons.

Bibliography

Al-Shemmeri, T. (2013) Engineering Thermodynamics. eBook, Bookboon.com.

Barclay, F. J.(1998) *Combined Power and Process: An Exergy Approach*, John Wiley & Sons.

Bejan, A. (2016) *Advanced Engineering Thermodynamics*, 4th edition, John Wiley & Sons.

Blair, T. (2016) *Energy Production Systems Engineering*. IEEE Press Series on Power Engineering, John Wiley & Sons, Inc.

Borgnakke, C. and Sonntag, R. F. (2017) *Fundamentals of Thermodynamics, Enhanced eText*, 9th edition, John Wiley & Sons.

Chandra, S. (2016) *Energy, Entropy and Engines: An Introduction to Thermodynamics*, John Wiley & Sons, Inc.

De Visscher, A. (2013) *Air Dispersion Modelling: Foundations and Applications*, John Wiley & Sons.

EIA (2013) https://www.eia.gov/outlooks/aeo/pdf/0383(2013).pdf

Feli x A., Farret, M. and Godoy, S. (2017) *Integration of Renewable Sources of Energy*, 2nd edition, John Wiley & Sons.

Heinsohn, R. J.(1991) *Industrial Ventilation: Engineering Principles*, John Wiley & Sons.

Hodge, B. K. (2010) *Alternative Energy Systems and Applications*, John Wiley & Sons, Inc.

Jones, J. B. (1960) *Engineering Thermodynamics: An introductory textbook*, John Wiley & Sons.

Manwell, J. F., McGowan, J. G. and Rogers, A. L. (2009) *Wind Energy Explained: Theory, Design and Application*, 2nd edition, John Wiley & Sons.

Moran, M. J. and Shapiro, H. N. (2006) *Fundamentals of Engineering Thermodynamics*, 5th edition, John Wiley & Sons.

Moran, M. J., Shapiro, H. N., Boettner, D. D. and Bailey, M. B. (2012) *Principles of Engineering Thermodynamics*, 7th edition, John Wiley & Sons.

Moran, M. J., Shapiro, H. N., Munson, B. R. and DeWitt, D. P. (2003) *Introduction to Thermal Systems Engineering: Thermodynamics, Fluid Mechanics, and Heat Transfer*. John Wiley & Sons.

Porteous, A. (2008) *Dictionary of Environmental Science and Technology*, 4th edition, John Wiley & Sons.

REN21 (2016) www.ren21.net/wp-content/uploads/2016/05/GSR_2016_Full_Report_lowres.pdf

Rhodes, M. (2008) *Introduction to Particle Technology*, 2nd edition, John Wiley & Sons.

Ristinen, R. A., Kraushaar, J. J. and Brack, J. (2016) *Energy and the Environment*, 3rd edition, John Wiley & Sons.

Conventional and Alternative Power Generation: Thermodynamics, Mitigation and Sustainability,
First Edition. Neil Packer and Tarik Al-Shemmeri.
© 2018 John Wiley & Sons Ltd. Published 2018 by John Wiley & Sons Ltd.

Sandler, S. I. (2017) *Chemical, Biochemical, and Engineering Thermodynamics*, 5th edition, John Wiley & Sons.

Stolten, D. and Scherer, V. (2011) *Efficient Carbon Capture for Coal Power Plants*, Weinheim: Wiley-VCH.

Williams, I. (2001) *Environmental Chemistry*, John Wiley & Sons.

Wiebren de Jong, J. and van Ommen, R. (2014) *Biomass as a Sustainable Energy Source for the Future: Fundamentals of Conversion Processes*, John Wiley & Sons.

Index

Conventional and Alternative Power Generation: Thermodynamics, Mitigation and Sustainability,
First Edition. Neil Packer and Tarik Al-Shemmeri.
© 2018 John Wiley & Sons Ltd. Published 2018 by John Wiley & Sons Ltd.